STREET RODDING TIPS & TECHNIQUES

HUNDREDS OF TECHNICAL TIPS ON ENGINE, CHASSIS, SUSPENSION, DRIVETRAIN, BODYWORK, ELECTRICAL AND INTERIOR FOR ANY STREET ROD PROJECT

FRANK ODDO

HPBOOKS

HPBooks
Published by the Penguin Group
Penguin Group (USA) Inc.
375 Hudson Street, New York, New York 10014, USA
Penguin Group (Canada), 90 Eglinton Avenue East, Suite 700, Toronto, Ontario M4P 2Y3, Canada
(a division of Pearson Penguin Canada Inc.)
Penguin Books Ltd., 80 Strand, London WC2R 0RL, England
Penguin Group Ireland, 25 St. Stephen's Green, Dublin 2, Ireland (a division of Penguin Books Ltd.)
Penguin Group (Australia), 250 Camberwell Road, Camberwell, Victoria 3124, Australia
(a division of Pearson Australia Group Pty. Ltd.)
Penguin Books India Pvt. Ltd., 11 Community Centre, Panchsheel Park, New Delhi—110 017, India
Penguin Group (NZ), 67 Apollo, Rosedale, North Shore 0745, Auckland, New Zealand
(a division of Pearson New Zealand Ltd.)
Penguin Books (South Africa) (Pty.) Ltd., 24 Sturdee Avenue, Rosebank, Johannesburg 2196, South Africa

Penguin Books Ltd., Registered Offices: 80 Strand, London WC2R 0RL, England

STREET RODDING TIPS & TECHNIQUES

Cover design by Tresa Rowe/Bird Studios
Cover photos by Frank Oddo
Editing, design and production by Michael Lutfy

First edition: May 2007

ISBN: 978-1-55788-515-9

PRINTED IN THE UNITED STATES OF AMERICA

10 9 8 7 6 5 4 3 2 1

CONTENTS

DEDICATION

To Robert J. Rider, my first street rodding buddy, and the guy who showed me the difference between a box wrench and an open-end.

THE COVER CAR: ROGER'S RIDE

Funny thing about the Ford cars for 1937. Those of us, who have been around, know that the 1937 models were, in all honesty, not particularly sought after in the mid-1950s. Oh, how times have changed. Now, I'm not going to sit here and tell you that Deuce roadsters and '34 Three-Windows are passed up in the mad rush for a 'Seven, but nowadays, they darn sure hold their own with their slightly younger post WWII brethren.

Another funny thing—of all the Ford V-8 annual production figures for 1932 through 1940, the 1937 count was the largest: 1,037,476! It was Ford's best year during the Depression.

Now that would lead the inexperienced to assume there is a 1937 whatever stashed away in every dark SoCal garage. Not hardly. In fact, Roger Fulmer of Brea, California, only found his in a Hemmings advertisement. And it was not exactly in his neighborhood. In fact, of the 33,683 DeLuxe Tudors that rolled out of Dearborn in 1937 with a window sticker of $580, he had to fly to Dallas, Texas, to consummate the deal! Roger knew what he wanted, and he went after it.

Roger's Tudor is the epitome of the mild-mannered, super reliable street rod that makes every like-minded hobbyist's pulse quicken. Let me tell you a little about it.

A closer look at the car reveals that it is not quite as stock as it may first appear. The body is completely shaved, with flush Honda door handles, and a Porsche fuel tank door thrown in for kicks.

The original 1937 frame was rehabilitated and boxed by Limeworks of Whittier, California. The independent front suspension and steering is a Total Cost Incorporated Mustang II styled assembly with rack and pinion steering. The rear is a Versailles 9-inch with parallel leaf suspension, and 3.50:1 gearing.

The engine is a basically stock 1989 Merc 302 with an Edelbrock intake manifold, 4V carb and polished rocker covers. Stepping away from the crowd in the search for reliability, Roger decided to stay with the stock cam and electronic ignition. The engine was built by Hail's Machine Shop of Fullerton. Keeping with the tractable theme, the transmission is a B & M shifted, rebuilt but stock 1989 Merc AOD.

The sheet metal was brought to perfection by Vintage Auto of Brea and then sprayed with Dark Cherry acrylic. The dark maroon leather upholstery was tailored by Modern Trim of Whittier over a stock Honda front seat and a much-modified Cadillac Seville rear seat. The rest of the interior is what might be expected for a gentleman of leisure—power front windows, A/C, VDO instruments, a custom console and a GM steering column.

No wonder Roger's grandson Keith Cousins looks the car over with unabashed awe. Don't you?

INTRODUCTION

The primary purpose of these pages, a companion to the *Street Rodder's Handbook*, is to pass on further (and sometimes re-emphasize) bits of practical information to the builder of, and "chief wench" on, a mainstream street rod. They have been selected for the fellow who labors day and night to first get his car on the road, and then (hopefully) at a more leisurely pace to keep it in top form. And too, when the temptation to upgrade becomes too great to ignore!

It is assumed that the typical reader is one who is familiar with basic automotive mechanics and tools, but more importantly, enjoys working on his street rod. He has to, for designing, building and maintaining a street rod is a hobby that consumes time and energy, to say nothing of money.

The first piece of advice is given in recognition of the novice rod builder's initial hurtle—determining the size and scope of the project into which he is about to throw himself. Sure, he knows that a full-boogie rework of an old car, or ground-up assembly of a reproduction requires many hours of garage time and operating capital, but he invariably underestimates the total! I call this tip my Rule of Three: No matter how long you think it will take, or how much you think it will cost, if you are going to do it well, multiply your best guess by three.

Failure to be realistic is why some projects never make it to graduation day. Therefore, you just might want to set your sights a little lower than an expectation of perfection. Of course it would be great to build and own a supercharged, chrome-plated, double-overhead-cam powerhouse mounted in a frame with state-of-the-art suspension, surrounded by a flawless body gorgeously painted and upholstered; in short, an award-winning showpiece featured on the cover of your favorite car magazine.

Instead, think about how much fun it will be to build and own a safe, reliable, reasonably comfortable old car that has enough moxey to scare off the rice-burning jelly bean next to you at the stoplight. Maybe your car is only primered, but it's got smooth sheet metal. Maybe the seat covers are a Wal-Mart Special, but they're clean. Maybe only driving it, and planning one mechanical or cosmetic upgrade at a time, would be more fun than just looking at a bunch of parts for a few more years. This may not be the stuff of which rod mag features are made, but man, it's Reality 101.

The second piece of advice I want to impart is also something I learned the hard way. It sounds deceptively simple . . . but it isn't. Decide what it is that you want to do with your street rod from the get-go.

Do you want an honest to goodness daily driver? I mean a car you can jump in every morning and go to work, rain or shine. (I know that concept may be outdated, but to my way of thinking, there is no more pure street rodding fun.) If that is your aim, remember a daily cruiser has a conservative powerplant; it idles smoothly and is easy on the gas. Sure, it has enough beans to hold its own in traffic, but first and foremost, it is reliable.

On the other hand, you may want a weekend cruiser, a slightly different breed of car. Typically, it is a rod with more horsepower than one really needs for transportation, but enough to provide its owner with his jollies at club meetings, rod runs, and the like where maximum tractability is not at the top of his want list.

Finally, there is the street rod that looks like a race car. (Or is it the other way around?) These are built by guys who want to do more than just strut their stuff, they want to get down and dirty in that much maligned "contest of speed."

I love 'em all! And, Lord knows, I've tried to have it all at one time or another during my life. As a consequence, I've learned the hard way; you can't have it all in one car. Like they say at the carnival, "Yer pays yer money, and yer makes yer choice." The point is . . . get started. There are a thousand things you'll have to do before you back the car out of the garage . . . and a hundred changes you'll want to make after you're up and running. Just keep things in perspective.

The third piece of advice is short and sweet: Promise yourself to complete at least one task a day, even if it's only to tighten a bolt. You'll get the car up and running with a minimum of downtime.

The fourth tip is the admonition never to forget one does not enter the street rodding arena for purely practical means. The man who wants practical transportation beyond the family sedan should buy a brand-new compact pickup and keep it for at least ten years. This is a hobby. As such, the money one spends is (or certainly should be) truly discretionary income.

Speaking of money, you older fellows don't need a book to remind you that is not the "pocket change" pursuit it was in the 1950s and early 1960s. Even so, the fifth tip is to look at commercial kits as a cheap way to buy time. Even if you have the skills of a professional machinist or welder, saving a day of personal labor for each $100 of the purchase price, puts you ahead of the game.

Relative to the tips I have selected for this book, I learned a long time ago doing the "Shop Manual" column for *Street Rodder* magazine, that I shouldn't impose my own tastes and preferences on those who wrote in with a couple questions. Admittedly, that is very difficult for someone as opinionated as I! Then, as now, I have tried to provide what I think are plausible answers to general problems. Nevertheless, honesty demands I also advise the reader when not to take on a do-it-yourself project. Discretion is always the better part of valor.

Beyond that, is the hands-on street rodder a relic of the past? Those outside the street rod realm might consider the ardent devotee to be out of touch with reality. After all, there are a number of fun-to-drive high performance cars in every dealer's showroom. Others might classify intensive rodding pursuits to be some form of obsession or addiction. Personally, I find such endeavors to be soothing. It's a labor of love, and those who do not know it, cannot appreciate it. Finally, I never would have entered this game nor stayed with it for more than 50 years had I not been so fortunate as to have a lot of helpful buddies along the way to give me advice as well as share the good times. I hope you can say the same.

—Frank Oddo

Chapter 1
ENGINE TIPS & TECHNIQUES

BONE YARD PICKIN'S

Crate motors are righteously popular, but hands-on rod builders still like to check out the inventory of the local automotive dismantler for a sleeper. Most yards have a selection of mills on wheels in the showroom where they are easily inspected.

But then there's the "Potter's Field" out back where accessibility is often at a minimum. Sure, check the odometer, but start your investigation with a thoughtful look at the type of damage the car suffered. You may well want to pass even if only a few thousand miles is documented. Read on . . .

Front-end wrecks generate destructive shock waves from the water pump back. Look for torn engine and transmission mounts, broken water pumps, cracked exhaust manifolds and ripped oil pans. If the force of impact got past the surrounding (and protective) frontal sheet metal, keep lookin' . . .

Rear enders result in longitudinal damages from the transmission output shaft forward. A cracked case is obvious, but transmission problems are often internal. And, as always, you want to buy the transmission along with the engine.

There's just no telling what went snap, crackle or pop in a T-boner close to the engine hold. A number of internal components may have been fractured or broken, and you

won't know for sure until each has been Magnafluxed.

Should you eliminate automotive dismantlers just because they have wrecks in 'em? No, of course not. Most cars got there because the cost of body repair made them a "total," not the condition of the engine. Just remember that automotive dismantler operators are true cherry pickers. When a new occupant is towed in, the manager carefully evaluates it to see if the engine should be pulled immediately for showroom display.

You can really get a good look at the engine in the showroom, but visibly damaged components are likely to have been replaced. You'll just have to use your powers of observation—gasket changes and paint touch-ups are dead giveaways.

When evaluating stock engines for that street rod project don't be confused by model years and advertised horsepower ratings in books such as Motor's *Auto Repair Manual*. (One of my favorite reference books.) "Advertised horsepower" was once determined by using an ideal dynamometer reading with no power-robbing accessories. These figures sound great, but are not realistic for an engine actually in service.

Conversely, don't be turned off by more modern Society of Automotive Engineers (SAE) ratings. These are an honest evaluation of an engine's power output measured with typical accessories attached. Stock compression ratios are as listed, however, and don't forget that most engines from the early 1970s on have low compression ratios. Of course, since octane ratings above 91 are a thing of the past, the lower ratios are more appropriate for street use.

Cast iron is tough. If you're looking at a pair of cylinder heads from a "burn job" that doesn't exhibit any significant scorching, they may be a great buy if the price is right.

ENGINE SWAPPING TIPS

This one is straight out of the "When Will They Learn Department." Most old cars have a smallish engine compartment that will accommodate a small-block of any kind much better than a big-block of any kind.

Sketch the dimensions of your engine compartment. Measure the width of the firewall and up front between the fenders where the original radiator once sat. Measure the distance between the two. With that information in mind, consider the prospective engine's dimensions. For instance, a big-block Chevy weighs 680 pounds as opposed to 580 pounds for the small-block. The big-block is 22 inches wide as opposed to 19.5-inches for the small-block. The big-block is 29.5-inches tall as opposed to 26-inches for the small-block. An inch here and an inch there add up.

The final index to a good engine swap is to determine that the crankcase/oil pan configuration comfortably clear the confines of the chassis with the intake manifold carburetor base level and the engine reasonably well-centered (side to side).

TWO LESS CYLINDERS

The admittedly rare pre-WWII Chevy body styles still draw enough enthusiasts to keep things interesting in a Ford dominated hobby. When the builder decides against a modern inline (as per the original mill) and finds the enormously popular small-block V-8 too big, a great choice is the carbureted Chevy 90-degree V-6 engine with its factory-mated, automatic transmission. It is a compact, lightweight package that will fit and look just right in a late 1920s Stovebolt. The engine weighs about 325 pounds, and the inexpensive 10-bolt Chevy rear, while marginal for V-8 torque output, is more than adequate for the V-6. Not only that, the factory-choice rear gearing would be in the ballpark.

A carbureted V-6 engine not only provides the obvious benefit of better fuel mileage; it also fits any early car's engine compartment with ease. Best

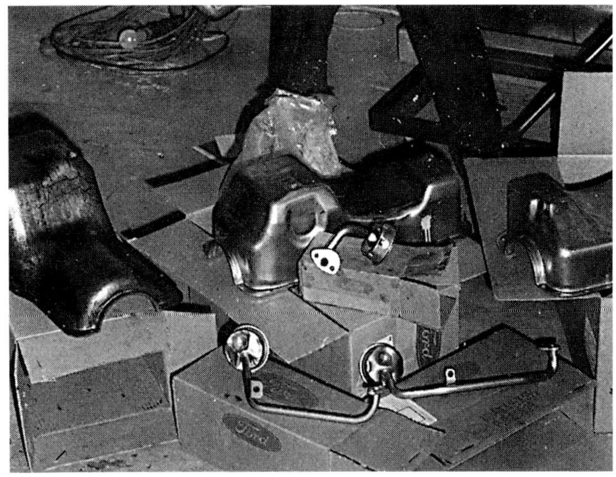

of all, with the proper drivetrain and gear ratios there's enough performance to guarantee satisfaction.

The GM 4.3L V-6 is almost the same size as a small-block V-8 except in length. In fact, it shares many parts with its V-8 brother. The distance from the water pump flange to the engine mount pad on the side of the block is the same for the V-6 and V-8. High performance parts are available, and it accepts a TH 350 tranny. Fit-wise, it's great for Model A rods, and the gain in fuel mileage is frosting on the cake.

Carbureted models of the American design Ford 3.8 liter V-6 continue to be of interest to economy-minded builders of the ever popular lightweight

reproduction 1927 "Track T" roadster. If the front sump oil pan clashes with your chassis and suspension components, there is a rearward oil sump option: oil pan E2TZ 6675A, oil pickup E3TZ 6622E, and oil pickup support bolt E2DZ 6345B.

PAN SWAPPING

The Ford small block, whether a V-8 or V-6, is not the easiest of swaps, but at least the company has provided a variety of oil pans from which to choose. A given dealer's stock is always changing, but today's computerized inventory (and a cooperative parts man) can locate what you need somewhere. Exhaust this resource before you head out to the automotive dismantler or ecology center for a fun-filled day dropping grimy oil pans!

NO MORE THAN FOUR

Any of the little inline 2.3-liter four-banger cammers Ford built from 1972 through 1984 with a carburetor and its factory-mated transmission (automatic for a simple installation, manual for better performance) makes a great engine swap in a

resto-rodded Model T or A. As the years wore on, the computerized and sophisticated adjuncts demanded by an ever-voracious EPA have added complexities that some rodders may not want to mess with, but visit a few automotive dismantlers in your area and look over the crop if you're interested in both fuel economy and pre-1935 four-banger street rod nostalgia.

A BETTER BUICK

For those experiencing clearance problems with the installation of a stock 3.8- or 4.1-liter Buick V-6, the solution is the water pump and front cover

assembly from the 3.0 liter V-6. The only drawback is that the engine-mounted fan can't be used. The part numbers for these pieces are as follows: Water Pump #25500407, Front Cover (which must be used) #25512585, Oil Pump Cover #25509135, Crank Pulley #25510369, and Water Pump Pulley #25510367.

FOOL 'EM!

Many flathead enthusiasts prefer the mystique of the pre-1949 heads (59-AB). But they also like the service friendly layout of the 1949–'53 block (8BA). A mix 'n match is possible; two holes in the block must be tapped for an eight-inch socket head pipe plug. Next, a water hole in the leading edge of the driver's side cylinder head (bottom photo) must be tapped for a quarter-inch socket head pipe plug.

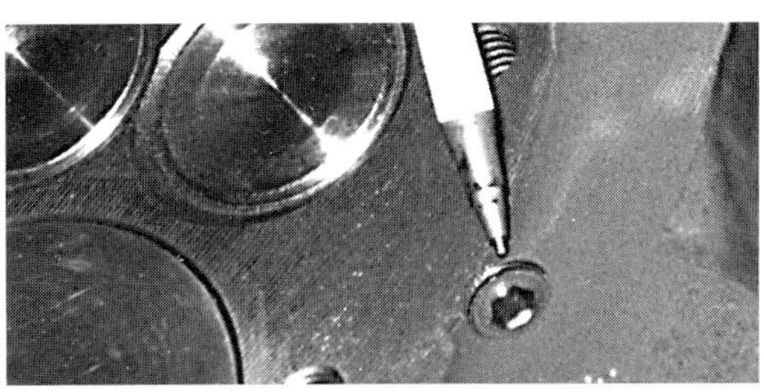

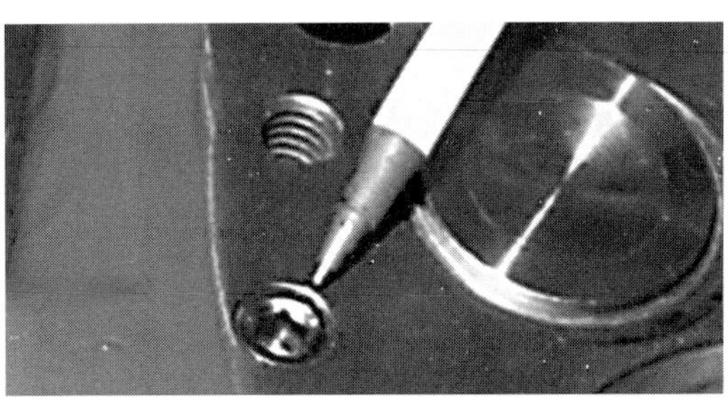

THE CLASSIC BOW-TIE SIX BLOCK

The rear of every Chevy six block (216/235/261 cid) is the same from 1937 through 1962, and the flywheel/crank flange pattern is the same from 1941 to 1962. Using stock Chevy parts, any engine from 1937 through 1962 can be installed in any 1937–67 passenger car or light truck chassis. Still need Chevy adaptation hardware? Check with Buffalo Engineering, 25625 Dahl Road, Arlington, WA 98223 (360) 652 7684.

Here are a few installation tips from Patrick's Antique Cars and Trucks, (P. O. Box 648, Casa Grande, AZ 85222; 520/836-1117). If your Chevy 235 has the big water temperature gauge hole, ask the local NAPA store for Balkamp PN 701-1542 or Weatherhead PN 20254 to use your original temp sender. The throttle linkage may need modification if the block's bell crank mount hole is further back toward the firewall. Take some of the curve out of the accelerator rod by squeezing it in a padded vise. It should clear the floorpan without the return springs attached. When that's done, measure how much it needs shortening by bringing the linkage assembly full throttle. Cut that amount out of the slender part of the rod end and reweld. The front engine mount pad(s) should be installed on the engine before putting it between the fenders since the passenger side bolt may only go into the mount with the timing gear cover and harmonic balancer removed.

A note on the front of the mount plates: They will interchange on every Chevy 216/235/261 built from 1937 through 1962. Also, the front and rears of every Chevy 6 are also the same as every small GMC 6: 228/236/248/256/270/302. Water pump/generator alignments depend on each other. Therefore, you must install the water pump first. If

your 235 came with the long water pump pulley, don't hack off the radiator core support or the latch panel. Simply contact Patrick's for the short pump and special pulley that allows you to use the larger, late-model fan. Patrick's can also supply the alternator bracket set. There's more, but you'll have to write Patrick's directly.

PICKING ENGINE COMPONENTS

It's one of those urges that the novice rod builder has to keep under control—choosing an induction system (cam and carburetor), a static compression ratio (via head milling or combustion chamber size), and a high-performance torque converter all

in the expectation of an rpm range that will hardly ever be reached on the street. Choose your engine components conservatively. You'll be much happier in the long run. A properly tuned, 2,500–3,000-pound street rod with a mild cam, appropriate rear gear ratios, and sensible driving won't deliver a knockout blow to a rodder's wallet.

CRACKERJACK

Several engine manufacturers put a prize in the crankcase—a forged steel crank that is dimensionally identical to the plain nodular iron number. One easy way to differentiate the two is when tapped on a concrete floor the forged crank emits a sharp ring rather than a dull thud.

FLAT-TOPS

Plain ol' flat-top (or dished) pistons, as opposed to the racy-looking domed variety) provide smooth, efficient combustion-flame travel and deter detonation.

CID MATH

Want to know how much of a displacement gain accrues from an overbored and/or a stroked engine? Use the following formula: Bore x Bore x 0.7854 x Stroke x No. of Cylinders.

FORD VS. MERC CRANKS

The 1949–53 Ford BA and Mercury CM crankshafts were improved designs over prior

models with slightly different counterweights. They also used insert-type rod bearings and came with a better main seal than the earlier, full-floating bearing 59A cranks. All of the 1949–'53 cranks

Displacement of 1953 Models
(Stock engines with stock .187-inch bores)

Ford 3.75-inch Stroke	Mercury 4.00-inch Stroke
(239 cid)	(255 cid)

Stock Torque Output of 1953 models

196 ft-lbs. @1800 rpm	218 ft-lbs. @1700 rpm

used the same seven-inch rod, but as is well known, the Mercury crank had a four-inch stroke while the Ford crank had a 3.75-inch stroke.

Flathead folklore has it that the Merc crank was also forged, but it is not. Both Ford and Merc cranks are high-quality cast-nodular iron. Even though they look the same to the casual eye, there is one simple way to cull the Merc four-incher from a batch of common Ford cranks. The 3.75-inch stroke Ford crank has a 3/8-inch plug in the journal. The Merc four-inch crank has a 5/8-inch plug.

Nevertheless, although flathead enthusiasts are charmed by the mystique of the Merc stroker crank, is the premium cost justified? You be the judge after checking out the stats in the box above.

As you can see, there's only a six percent increase in displacement, and only an eleven percent increase in torque production.

SONIC TESTING

Most Ford enthusiasts are aware that when it comes to the late model OHV 289/302/351 Windsor "Thin Wall Castings," an overbore greater than 0.030-inch is verboten. Chevy V-8 enthusiasts have it better, but one should never get greedy. If you want to safely maximize an overbore with any make block, start looking for a machine shop that has the ability to sonic-test that hunk of cast-iron. No block slated for high performance should be less than .200-inch thick on the "thrust" sides of the cylinder banks—inboard on the driver's side and outboard on the passenger's side.

A FINE BALANCE

It's fairly well known that precision balancing an engine's reciprocating and rotating components will make it run more smoothly and live longer. High rpm vibrations will prematurely end service life. The difference between an internally balanced engine and one that is externally balanced can often be confusing, however.

When the crankshaft and piston assemblies are balanced as a unit, and the remaining components that are attached to the crank, i.e., the flywheel, clutch assembly and harmonic balancer, are individually balanced for additional smoothness the engine is said to be "internally balanced."

When everything that is attached to the crankshaft is balanced as a unit, the engine is said to be "externally balanced."

Ford small-blocks are externally balanced. When sending the rotating and revolving parts attached to the crankshaft out for balancing, don't forget to include the flexplate.

A small-block Ford can be converted to internal balance by plugging the crank counterweights with "heavy metal," usually tungsten. However, it is relatively expensive, and ordinarily restricted to high-performance engines.

Porting and relieving a stock flathead block is about as traditional as the venerable old mill itself, pro and con opinions of performance gains not withstanding. Porting requires a die grinder, and beyond the need for a deft touch, it isn't brain surgery. Relieving the head surface of the block isn't quite so easy. A router equipped with a cylindrical, flat-bottom grindstone (available from any industrial hardware store) makes the job a cinch. The flathead block is a perfect table for the router plate and the infinite depth adjustment makes

consistency a snap. Spray the block with paint (you'll need to have it hot tanked later, anyway) and use a template to define the perimeter of the cut. The result is a perfectly smooth "engine-turned" finish.

HEAD HUNTING

Rebuilding that veteran street rod engine that was designed for high-octane gas? The most practical way to deal with today's pitiful excuse for "premium" is to lower the compression ratio by switching to compatible cylinder heads with open (larger) combustion chambers. They are available for most popular makes and models.

Even though head milling is a traditional way of providing a compression boost (something we are trying to avoid) moderate to mild milling insures a perfect mating to a freshly surfaced cylinder block. When you bring your small-block Ford heads to the machine shop, however, also cart along the intake manifold you plan to use. The shop will also have

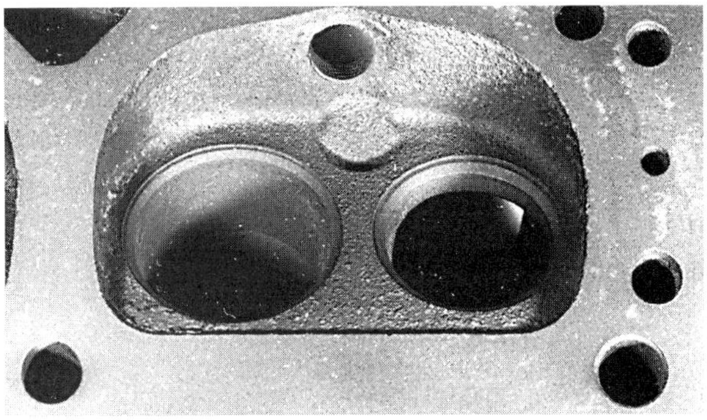

to mill the bottom of the manifold and the intake port interfaces. Any competent shop is aware of this, but it doesn't hurt to verify.

STUD FARMING

In a high-performance short-block rebuild, using studs, hardened nuts and washers instead of bolts to fasten the cylinder heads and main caps is the only way to go. The cost is easily justified, however, even if you're only building a strong street motor.

Studs prolong thread life in any number of components and minimize distortion by distributing torque forces evenly. When directionally appropriate, studs also provide a helping hand aligning gaskets, and especially so when installing V-8 flathead lids!

Also, bolts should not be used where the original equipment manufacturer used studs. Such is the case with Ford's small-block intake manifolds. The OEM studs on the corners prevent the gasket from slipping and sliding.

STEP TO THE FRONT

Not all, but many aftermarket head gaskets are stamped "FRONT" on one end to insure correct installation. The notation is there for a reason—such gaskets switched end for end can interfere with coolant flow. By the way, unless head gaskets are definitely marked not to apply any stickum, a thin

coat of Copper Coat Gasket Compound (as per directions) is worthwhile.

FREEZE PLUGS

Home-bound engine builders are often frustrated when it comes to installing those attractive brass freeze-out plugs—they always want to shift in the block. Look for a large half-inch drive black impact socket that fits snugly in the cup. Insert a short ratchet extension and gently tap them in with a brass hammer.

CAM GRINDS

We all love the *rump, rump* of a hot cam, but high-performance cams typically fall short of good throttle response at low rpm. And, the truth of the matter is, that's where we drive most of the time. Study the cam catalogs carefully, and stay with the recommended milder grinds.

CAM BREAK-IN

Always break in a new cam with the engine speed varying between 1500 to 2000 rpm for 20 to 30 minutes. Moreover, a new camshaft and valvetrain always requires a special lubricant. Isky's "Rev-Lube," for instance, contains moly-sulfide and zinc dialkyl-di-thio-phosphate. Whew! It should be applied liberally on the side of the lifter, on the tips of the push rods and rocker arms and on the valve stem ends.

Why is it necessary to take such elaborate precautions? Without this high-temp lubrication, the cam lobes and/or other valvetrain components could fail. The resultant debris could easily contaminate the entire engine. 'Nuff said?

A STITCH IN TIME

By the way, the special lock nuts used on stock rocker arm studs don't last forever. If you are planning to reuse the originals because they still look good, don't. Replace them.

LIFTER ADJUSTMENT

Hydraulic lifter adjustment (pre-loading) can be a problem for inexperienced street rodders. On engines with non-adjustable valvetrains, torque the

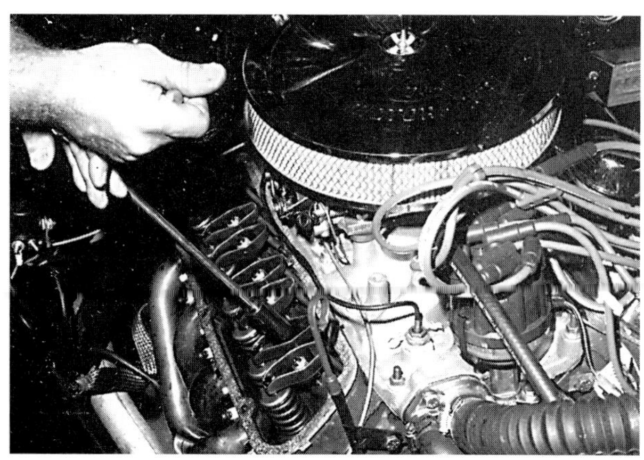

adjustment nut to recommended specs. After you fire the engine, back off each nut until the rocker begins to clatter. Tighten just enough to stop the clatter.

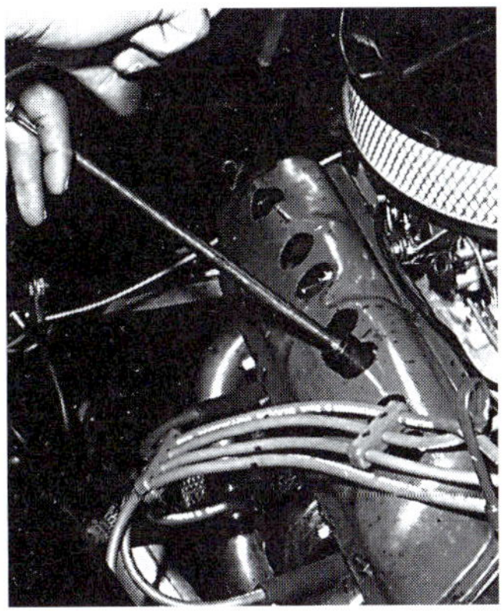

DON'T LITTER

Adjusting hydraulic lifters can be quite messy. The speed shop where you got the cam can order rocker arm "baffles" that reduce excessive oil flow from the lifter valley. Or, you can make an inexpensive "tool" for the adjustments. Cut holes in an old tin rocker cover just large enough to slip a deep socket through to the nut.

THE SOUND OF SILENCE

After you've racked up some mileage on the engine, you may hear a clatter in your rocker arm assembly. If you aren't certain of the exact rocker, grab a large screwdriver for use as a stethoscope. You should be able to pinpoint the offender(s) before removing the rocker cover. (Make sure you put your ear on the handle . . . just kidding!)

A BANK JOB

Performance-minded street rodders who choose a mechanical over a hydraulic cam must remember that periodic valve lash adjustments have to be made with the engine at the normal operating temperature of 180-degrees. And it is best to work on only one bank at a time. Pull the rocker cover, and verify each lifter to be adjusted is on the base circle of the cam by wiggling the rocker. Adjust the exhaust valves first, When that bank is done, replace the cover and restart the engine. Give it time to again reach operating temperature, then shut it off, pull the cover on the other bank and adjust those valves.

RACK 'EM BACK

If you ever have to remove (non-roller) valve lifters, each must be returned to its original bore. When the lifters and cam were new, each lifter developed a unique wear-in pattern, and if they get mixed up, subsequent wear will significantly shorten their service life. Make a rack and number the holes for intake and exhaust as well as cylinder.

THE HOLE STORY

If cast-aluminum rocker cover breathers are your thing, remember they should be mounted as high as practical to avoid leakage. Four or five eighth-inch holes drilled through the rocker cover below the breather are all that is needed to provide adequate crankcase ventilation.

FOR THE TIME BEING

Wish you had a precision degreed harmonic balancer, but don't want to pay the machine shop to index the balancer just yet? A carefully applied timing tape installed properly at the factory inscribed "0" on a clean balancer will do the trick for the time being. Eventually it will become soiled, so get your local machinist to index the balancer permanently when you can.

CARBURETION TIPS

Fuel injection is well on its way to becoming the choice of induction for high tech street rodders, but the majority of hands-on rodders still like conventional carbs atop their engines. A glance at any current manifold company catalog will ascertain that traditional street rod induction is far from obsolete. A closer look may prove confusing, however, since there are so many varieties. Generally

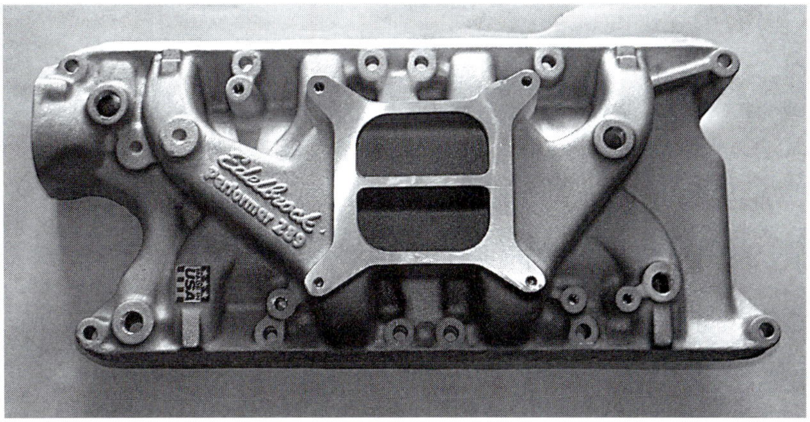

speaking, factory intake manifolds have an advantage in economy, whereas aftermarket manifolds have the advantage in low and mid-range street performance. A quick overview should prove helpful.

The dual-plane manifold incorporates two separate runner cavities, one for each half of the carburetor. This design allows the fuel/air mixture pulses to pass alternately from one plane of the manifold to the other in the firing sequence. It is an efficient arrangement for mild street engines.

A single-plane intake, other the other hand, connects all runners to a common plenum and delivers improved cylinder-to-cylinder air/fuel

distribution because of its straighter and shorter runner lengths. Rodders with a more potent engine prefer them because they are strong in the 2,500 to 6,000 rpm range.

Heat riser passages (those often darkened midpoint areas on both sides of a stock manifold) utilize exhaust heat to help keep the fuel in a vaporized state. There is a downside, however. The heat decreases air density with a slight loss of power. Performance manifolds without heat risers ("cold" manifolds) can decrease gas mileage enough to make you think twice about installing one.

THE SPACE RACE

Carburetor spacers aren't designed to lift the carb up in a crowded engine compartment. Four-hole spacers provide better low-rpm torque/throttle response on intake manifolds that may be a tad too "racy" for the street. Or, if you like that manifold, but want to downsize the carb and increase the fuel mixture velocity, use an open spacer. It will provide the increase in carb-to-plenum floor distance needed to increase high-rpm power where more plenum volume is required.

Regardless of manifold design, poor fuel economy is usually the consequence of excessive carburetion.

That said, it must be pointed out that over-concern with economy is self-defeating when it comes to street rods. After all, you didn't get into this hobby to save money.

JUST PASSING THROUGH

The term cubic feet per minute (cfm) is a measure of how much air can pass through the carburetor at a given vacuum drop. It is determined by the design of the carburetor, not the size of the

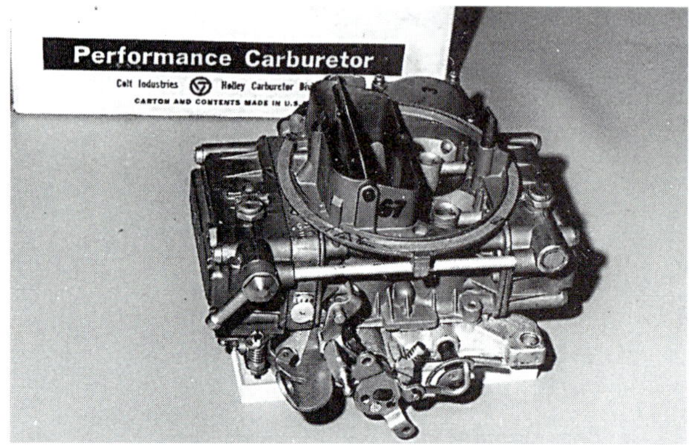

jet. Four-barrel carburetors with "vacuum actuated secondaries" restrain the carburetor, inasmuch as they only open on throttle demand. Nevertheless, the wise street rodder will install a carb with the smallest cfm rating he can get away with to avoid the common pitfall of excessive carburetion.

To determine "what you can get away with" for a mild street engine, multiply the engine's cubic inch displacement by two, and then subtract ten percent. It'll keep you out of trouble every time.

EVERYTHING OLD IS NEW AGAIN

Then there are the traditionalists that roam the swap meets looking for good deals on used high-performance intake manifolds. Old intakes can be bead-blasted to look like new. Well, almost new.

Unfortunately, one can't just eyeball an old manifold to know with assurance that it isn't warped. However, if the price is attractive enough to take a gamble, you can determine if it will adequately seal when it interfaces with the heads, without installing carbs and firing the engine. Pick up a set of fresh gaskets and lightly coat the tops and bottoms with white (Lithium) grease. Torque the manifold down just as you would for real. Then unbolt it and visually inspect the crush imprints in the gaskets around the port areas.

If you find any reason to doubt the efficacy of the seals bring manifold and gaskets to an automotive machine shop. An experienced eye can tell if the manifold can be trued on the mill.

THREE FOR THE SHOW

You've been thinking about giving the ol' faithful 4V a rest and decorating the topside of your Chevy small-block with a triplet of chrome-capped pots.

You know that Edelbrock still catalogs a brand-new three-deuce manifold (C-357-B) and you've seen used manifolds at swap meets. Carburetor-wise, Edelbrock's catalog recommends the small 1-7/16th bore Rochester 2GC, but where do you go from there?

The decades old two-barrel 2GC Rochester carbs ("C" for choke) came in two varieties. The first was fitted to innumerable plain-Jane GM small-block intakes beginning in the mid-1950s. And, indeed, it was the center carb on the subsequent "Tri-Power" arrangement that appeared on the 1958–'60 Cadillac El Dorado, 1958–'61 Chevy 348 c.i.d., 1958 Olds J-2 and 1958–'60 Pontiac. Occasionally one of these set-ups momentarily appear along the swap meet trail. Get your wallet out fast if you spot one!

The outboard carbs (2G), however, did not have chokes or idle circuits. When the three are jetted according to displacement and fitted with mechanical, progressive linkage, they make an excellent traditional street combo that not only looks great, but runs very well.

Rebuild kits for the conventional Rochester 2GC are available, as are rebuilt carbs (on an exchange basis). Finding the dedicated Tri-Power outboard 2G carbs, however, will take some looking. No problem, though, simply close off the idle circuits in two more 2GCs and use them as outboard carbs.

CARB MYSTIQUE

Limited availability notwithstanding, few carburetors can hold a candle to the Stromberg 97 (left) for flathead traditionalists.

If all you remember about them is that they were leakers, it's because you didn't run a fuel pressure gauge set at two lbs.

Stromberg 97 rebuild kits have been offered by Speedway Motors, (P.O. Box 81906, Lincoln, NE 68501; 402/474-4411) and Patrick's. However, only they know what's currently in their inventory. Give 'em a call.

Another traditional two-barrel carb (left), and one for which kits are more readily available is the Ford/Holley 94. It is my personal favorite these days (I love the Ford script) and three of them are slated for the Ford 302W in my own coupe.

If you're wondering how to mount these or any other carb on whatever mismatched manifold you have, rest easy. Adapter plates are available from Speedway and Offenhauser Sales Corp. (P.O. Box 32218, Los Angeles CA, 90032, 323/225-1307).

Indeed, Offenhauser has a carb adapter plate for just about any mismatched intake manifold and carb you have.

Progressive linkage kits (or at least the bits and pieces) for three carbs in a row may be languishing in your local speed shop, but both Patrick's Antique Cars and Trucks and Speedway Motors have cataloged them.

THINK OUT OF THE BOX

Got an awkward angle on your throttle linkage? Don't bend the rod by hand, use your tube bending tool for a perfect "90."

CALL AHEAD

Oh yes, with regard to businesses that deal partially or fully in street rod nostalgia, parts (both new and used) come and go. It never hurts to telephone first, but always followup with a letter reiterating your needs. (Telephone calls are only for instant information, the gist of which is quickly forgotten by the best of us.)

GASKETOLOGY

High-quality intake manifold gaskets are as important to efficient engine operation as head gaskets even though they take much less "environmental" abuse. I've had better luck with thick fiber-type gaskets than the tin types.

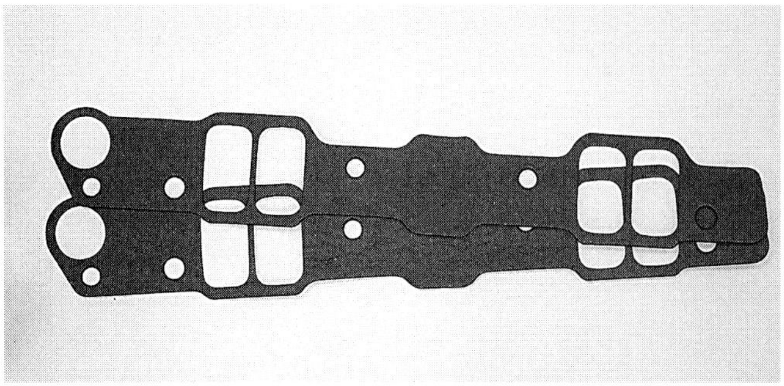

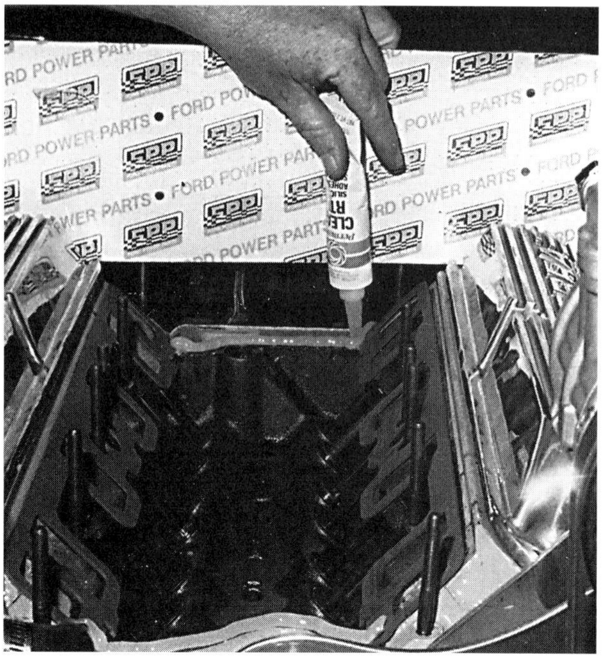

When installing an intake manifold and you've misplaced the cork end gaskets, use silicone in their place. It works as well or better than cork or neoprene types. And, if you're not certain about the tightening sequence, work from the inside out, and torque the studs and bolts to 23–25 lbs.-ft. Loctite Threadlocker on the washers and fasteners adds a touch of permanency.

JET LAG

Regardless of the carb combination you choose, jetting is something you have to consider. If you are going to run a three deuce set-up, however, start with just one 2V carb in operation. Note which jets are factory installed, but don't change them yet. Block off the outboard carb bases on the manifold with aluminum plates.

If the engine backfires while it is warming up, chances are your fuel mixture is too lean, and will not ignite properly. Before fiddling around with the adjustment screws, look for a manifold leak. It can sometimes be hard to locate, but typically there is a missing plug or cap, or less commonly, a bad gasket.

If the engine idles too rough, too slow or too fast on start-up, the likely cause is a maladjusted idle mixture. The fix is to adjust each idle mixture screw to deliver the highest vacuum reading with the assistance of a hand-held vacuum gauge.

When you feel comfortable enough to take the car out for a shake-down run, buzz it up. If you notice a "flat" spot while you're accelerating, it's probably the jet sizes. Pull over to the curb, and run the engine up to a steady 2,000 rpm while you're sitting there. If it loads up, it's jetted too rich. If it hesitates, chances are you're jetted too lean.

Install a set of new spark plugs, and "read them" after a hundred road miles. If the engine performs smoothly up to about 3,000 rpm in high gear, and the plugs exhibit neither rich (sooty black) nor lean (glossy yellow) coloring, chances are good you are in the right ballpark.

Now you can jet the outboard carbs 0.010-inch under whatever is in the center carb and install them with the progressive linkage. If after another 100 road miles your plugs still look good, and your tail pipes are colored ash gray to light brown, you are right on for what, and the way, you drive. If not, lean out or fatten up the outboard carbs accordingly.

MORE POWER TO YOU

Supercharged street engines have their own idiosyncrasies, and if you've mounted a Holley carb, one of them is blowing out the power valve on

start-up. This often happens when the throttle is opened too abruptly and the engine backfires. Before the ignition key is twisted, two quick pedal presses will prime the accelerator pump

To some extent the same holds true for less aggressive street rodders who mount the ever-popular Holley on a mild, normally aspirated small-block. Sometimes a backfire will blow out the power valve. The company does offer power valve check ball kits for a number of models, however. The best place for up-to-date info is your local speed shop.

FEED THE FIRE

Always install an electric fuel pump as close to the fuel tank as practical, naturally keeping in mind the proximity of exhaust pipes.

While you're down there, and the tank is empty or out, install an inline pet cock. I guarantee it will come in handy some day. Another good idea is a manual fuel shut-off toggle switch in a discrete location.

Street rods often sit for days at a time, and since most of us use electric fuel pumps, it's a good idea to relax for 15 seconds after turning the ignition on and before you spin the starter. You want to be sure the carb is "primed."

Richard Lazzeri of Quincy, Illinois passed along this recommendation for Chevy V-8 rod builders—use an AC /Delco mechanical fuel pump (PN 6471180, PN 41216) in your next project. Chevy installs this pump and others like it, in Corvettes and trucks. It has a fuel-return outlet that permits the installation of a quarter-inch fuel line back to the tank. By returning fuel to the tank, the fuel in the pump is constantly being replaced with cool fuel. The carburetor is also getting cool fuel that helps it run better in hot weather. Richard reports that the pump works great in his 1937 Chevy street rod.

IGNITION TIPS

The first time I personally experienced glowing red-hot headers I was, admittedly, a bit unnerved. I was breaking in a fresh engine, and even had a garden hose trickling water into the slowly overflowing radiator. The cause turned out to be retarded ignition timing. That should be the first thing to check if it ever happens to you.

Just what constitutes a street rodder's "non-stock" engine is an open question. And so is the exact ignition timing where a modified engine runs best. After you hook up your timing light, try this—advance the initial timing until the engine pings under full throttle, and then back the timing off two degrees.

By the way, swapping out cast-iron exhaust manifolds for headers can also cause a lean mixture condition. In that case, it should be corrected by carburetor adjustment, otherwise the valves may overheat.

Those transparent distributor caps are pretty, but the downside is that they are more prone to carbon tracking and do not insulate as well as stockers or aftermarket caps made with alkyd resins.

Ford Roller Distributor Gear—Once upon a time the drive gear on a small-block Ford distributor was a "no brainer" unless you were running a steel billet cam and needed a bronze gear. However, since factory roller cams were introduced, matters have become more complicated. The drive gear chosen must match the drive gear on the cam, or both will quickly be history. There are a number of varieties, and when in doubt, your first resource is either the cam grinder or the manufacturer of the distributor. The parts man at the local Ford dealer can help with all stock items.

Unfriendly Fire—Everyone knows that the correct firing order is essential, and that mixed up primary wires can create a whole host of problems including severe engine miss, spark knock, and surging. Sometimes, however, these pesky problems are caused by what is called "induction crossfire." This occurs even though the firing order is laid out correctly, but the plug wires are not routed just right in the looms. For instance, the Ford small-block firing order is 1-5-4-2-6-3-7-8, and even though the wires may be strung out properly from the cap to the plugs, if #7 and #8, or #2 and #4 plug wires are "loomed" next to each other on the rocker cover, induction crossfire can occur. The best way I've found to make sure everything is in order is to install those handy little numbered clips on the plug wires at the cap and also down where the plug wires come out of the loom.

Here's another flash from the Bad News

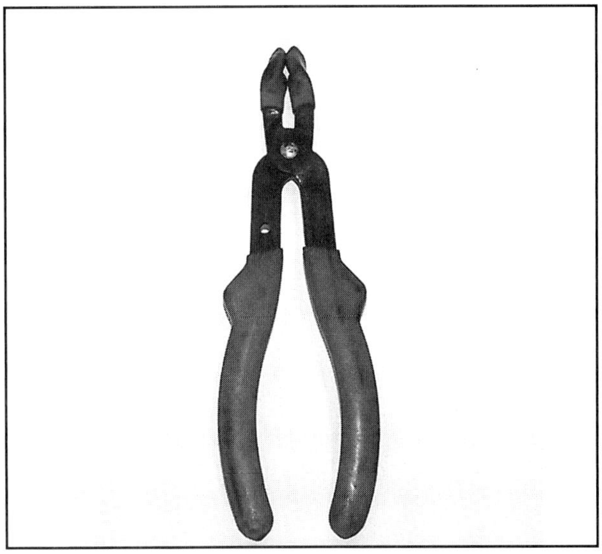

Department: If you get one or possibly two plugs next to one another with rusty deposits on the insulators, you can be sure of water in the cylinders. That's not good; you either have a blown head gasket, a cracked head or a cracked cylinder wall!

A compression gauge is a mighty handy tool to own for diagnosing the above and a number of other problems. Reading the gauge is no problem, full instructions come with the kit. What the instruction folder won't tell you is that the conventional gauge (as pictured) can be a problem for flathead enthusiasts. The threaded portion of a screw-in type compression gauge is often about 1/8-inch longer that the threaded portion of the spark plug. If it is used on a flathead, it is quite possible the exhaust valve will hit it and bend. The old-style hand-held, rubber-cone–type gauge that is forced into the spark plug hole is best.

By the way, sometimes you just have to pull a spark plug when the engine is hot. A tool that may not be in your toolbox, but comes in mighty handy in such cases is a spark plug wire puller. Just don't pull though, gently twist the insulator left and right until it comes off.

A SPARK PLUG PRIMER

The mantra for a tune-up used to be "plugs 'n' points." Well, points distributors have almost disappeared from the scene, but not spark plugs. Yet, if you drive a late-model grocery getter, they do seem to last forever. Modern electronic ignition and fuel systems are why, but the street rodder shouldn't become blasé. Even in these days of high energy, multiple discharge, magnetic reluctance and other electronic wizardry, the higher-than-ever combustion chamber temperatures still contribute to spark plug tip erosion. That's why it's a good idea to follow a more or less traditional spark plug inspection and maintenance schedule to assure peak performance and gas mileage.

Beyond that, does Rodney Street ever consider the fact that his plugs are designed to bask in 600–700 degrees F while the engine is idling? That's not particularly hot, either; just warm enough to prevent fouling. Furthermore, the tip temperature will average about 1,100 degrees when cruising, and when he gives it the gas, temperatures at the tip of the plug are well over 1,200 degrees. (They must not exceed 1,600 degrees or the fuel will ignite before the spark occurs, a destructive condition called "pre-ignition.")

Heat Range—Most manufacturers identify spark plugs as "hot" or "cold" with a numbering system: the higher the number, the hotter the plug. The manufacturer will tell you which heat range is "normal" for your family car's engine, but because your street rod engine is likely to be modified, the rest is up to you.

If your engine has built up an appreciable mileage, or city driving is the rule, a one-step hotter plug (over factory recommendation) may be necessary. However, considerable highway driving may require a one step colder plug. Sometimes, especially in engines with increased displacement, higher compression, multiple carburetors, a high-lift cam, etc., you may find you need two sets of plugs—a hotter one for city driving, and a colder one for sustained highway use. It's all part of the price you pay for hopping up your engine.

If, however, your displacement, valve timing, and compression ratio is close to stock, and your driving is equally distributed between city and highway speeds, you should be able to use the factory recommended plug. Nevertheless, if you think you may be having problems, or you are just interested in plug experimentation, you have to understand what a spark plug's heat range entails and how to read your plugs.

A "cold" plug does not have any measurable effect on the combustion temperature in the cylinders, but it transfers combustion heat from the electrode and insulator to the cylinder head and surrounding water jacket more rapidly and effectively than a "hot" plug. The hot plug retains more heat within the plug itself between firing impulses.

Technically, the plug's heat range is a measure of its ability to conduct heat away from the firing end, and remain in "thermal equilibrium." (Thermal equilibrium is the ideal condition where heat conduction from the tip of the plug to the cylinder head increases as the plug gets hotter until heat loss is equal to the heat gained from the burning fuel.)

Cold plugs use short, stubby insulators to quickly transfer heat. Hot plugs have long insulators that do not carry away heat quite so fast. On "extended tip" plugs, the tip projects more deeply into the combustion chamber, and is cooled by the incoming fuel/air charge thereby broadening the heat range.

Although cold plugs foul more easily when cruising around town, they won't damage the engine. Plugs that are too hot for freeway driving, though, can result in engine damage as pre-ignition can soften the crown of a piston before a loss in power is noticed. The very essence of street rodding—the mix and match of automotive components with far more than mere transportation in mind—defeats the "one size fits all" concept. With that said, let's see what we find when we pull the plugs . . .

Reading Plugs—Reading sparkplugs is an art. Begin your analysis with new plugs in the "best guess" range (as per the manufacturer's instructions for the year and model engine you run). Drive normally for 200 miles or so, and then remove the plugs. A word of caution here, never pull on the spark plug wire itself. Grasp the rubber boot of the plug wire and twist it from the plug. If you have a cramped engine compartment, you may want to stop by our favorite auto parts store and buy a special pair of pliers for this.

Beyond that, make sure there isn't any dirt in the plug well that might fall into the cylinder during removal or installation of the plug and always use a designated spark plug socket incorporating a rubber sleeve to avoid cracking the insulator. Tag the plugs or line them up in the order they were removed. You want each plug to identify the conditions existing in the cylinder from whence it came.

Reading a spark plug is essentially an analysis of the color of the porcelain insulator. Study it with a magnifying glass, in the sunlight if you can. It takes time for a fresh plug to color (that's why you first put several hundred miles on them), but if the heat range is correct, the insulator will turn a light chocolate brown. On the other hand, if the plug is too cold, it will be black and oily in appearance. If you find small deposits of contaminants on the insulator, you may be running an incorrect heat range or you may have a poorly adjusted carburetor.

There is a significant difference between plugs that are oil-fouled and those that are carbon-fouled. Oil fouling is black, wet, and sticky whereas carbon fouling is black, sooty and dry looking. The latter indicates an overly rich air/fuel mixture, a weak ignition or simply a plug that is too cold.

In a worst-case scenario, where the center electrode and ground electrode look like they are starting to melt, you probably have over-advanced timing and a lean air/fuel mixture in addition to a plug that is too hot. Make some changes quickly. If you want backup support relative to your judgment, compare their color and condition against the color charts provided by plug manufacturers and available in speed shops and

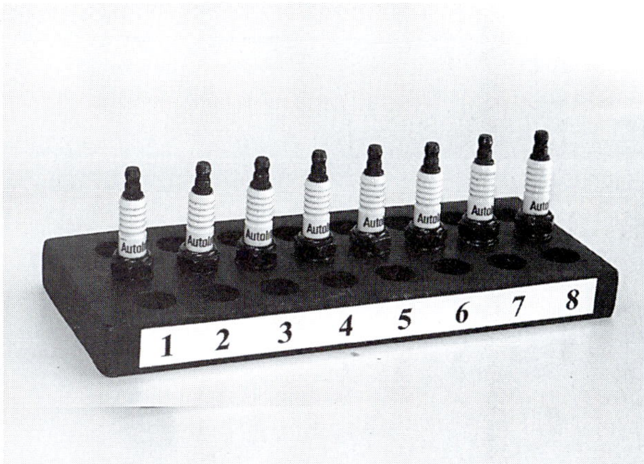

auto parts stores.

Real sticklers for engine condition and performance keep a "visual database" along with the written notes in their street rod log book. Drill three rows of four, six or eight (depending on the engine), in a small piece of wood. The old plugs go in the first row in a numbered sequence and in the condition they came out of the engine for careful evaluation at a more leisurely time. A new (or freshly cleaned set) go back in the engine.

The first sign that a plug is too hot is porcelain that refuses to color or has a burned, glassy (high-speed glazing) or perhaps a yellowish, varnish-like color on the insulation. In extreme cases the electrode will turn blue, and if that is happening in your engine, you are on the ragged edge of disaster—detonation and pre-ignition.

If your engine is "going away" you'll find an oil deposit on the plug. Oil is leaking past worn valve guides or piston rings into the combustion chamber. Fix the engine, not the plugs.

Plugs that show a brown or grayish tan color and only slight electrode wear are the correct heat range for the engine and operating conditions. Plugs that are worn out will also appear normal in coloring, but the electrodes are rounded. Discard them.

Dry carbon deposits indicate a rich mixture or a weak ignition. A slightly hotter plug will compensate temporarily, but the problem(s) should be corrected.

Cleaning—Cleaning used, but not damaged spark plugs may seem archaic to folks with a "throwaway mentality." Nevertheless, there are more than a few of us (myself included) that still adhere to the philosophy of "waste not, want not." I keep a set or two of low mileage plugs that have been cleaned and gapped and ready for periodic changes. I discard them after an accumulated 12,000 miles of service life. (Yes, I keep a logbook for my street rod as well as for my race car.)

Spark plugs should be cleaned with a compressed air blaster filled with a special abrasive, not with the wire brush wheel on your bench grinder. First, however, wash them in clean solvent and dry them off. (There are two reasons behind these admonitions: a bit of the abrasive could stick to oil residue, and wire brush wheels leave metal streaks on the insulator that could cause a short-circuit.)

Don your protective eyewear, and insert the plug in the blaster. Rotate it between your fingers and rock it back and forth to insure full cleaning. Most blasters can be switched over to compressed air only in order to blow away any remaining abrasive. Do not destroy the plug with prolonged blasts in the same spot. Gently roll it around on the rubber seat to evenly distribute the abrasive blast. Be careful not to cock the plug so much that you get a face full of sand. It hurts.

Gapping Plugs—Spark plugs in the correct heat range will operate efficiently as long as they are clean and properly gapped. But the kicker is, which of the several types of gapping tools available is the best to use? There is only one aspect in which credible resources agree: A precision gap is less likely with a blade-type feeler gauge, particularly on a plug that has been in use for a time because of electrode erosion.

Admittedly, some manufacturers do not recommend pliers-type gapping tools. Sorry 'bout that, my favorite gapping tool is an old-timer from Snap-on Tools (CF-70A) that insures a perfectly precise squeeze with no danger of over-bending the electrode or cracking the porcelain. But it is hard to fault the round wire gauge you can buy everywhere for a couple of bucks or a ramp type on a key chain that you can get for free!

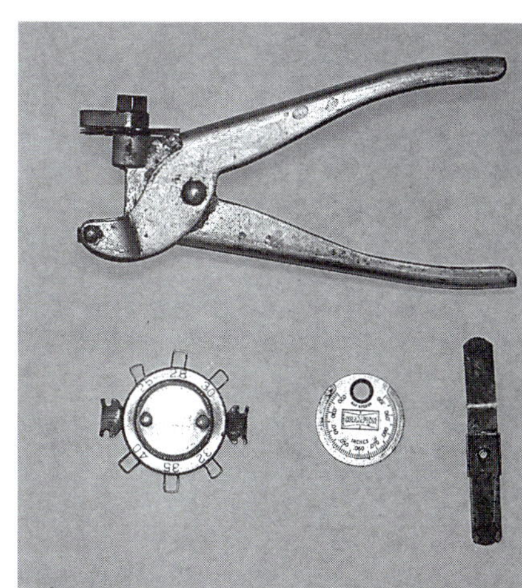

Generally gap your plugs to

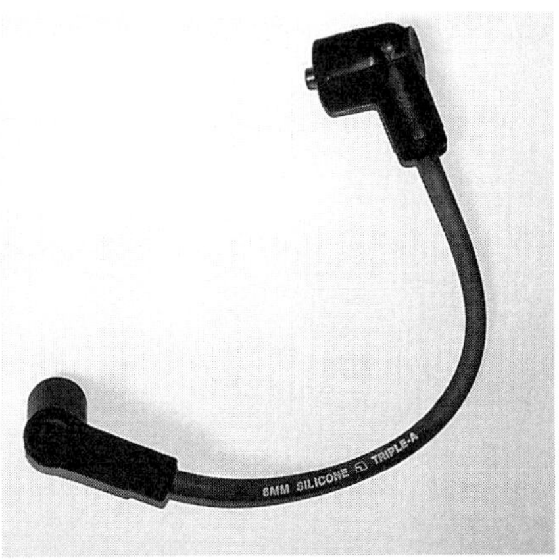

the mid-point of the range given in the manual. A gap that is too small will cause rough idling, excessive fuel consumption, uneven power, and all 'round poor performance. If the gap is too large, you will experience misfire at high speeds.

If your engine uses gasket-type plugs, always use fresh gaskets. Screw the plugs all the way in finger tight.

Noise Suppression—When a spark plug fires, high-frequency electrical oscillations that cause radio interference are set up in the spark plug leads. Resistance-type plugs for noise suppression are superior to carbon-core (TVRS) spark plug wires and won't deteriorate as rapidly.

If you still detect radio interference, install a piece of resistor wire from the coil to the cap. But resistance-type plug leads and resistor spark plugs should not be used together. Their combined resistance can be high enough to impair ignition system performance.

Copper-core leads do not deteriorate internally although the exterior insulation will eventually become brittle. Just remember to maintain sufficient spacing between the wires to avoid the possibility of cross firing due to mutual inductance.

It's true that sparkplug analysis can become a "heady" science, but few components are subjected to such a beating. Pay attention to them. They'll tell you plenty.

STARTERS AND FLYWHEELS

Here are a few pertinent considerations about starters and flywheels . . .

Never grind on the starter for longer than 10 or 15 seconds, and wait at least that long before trying again to avoid a burn-out. It is an old admonition, and newer, high torque starters are more durable, but it's better to be safe than sorry.

A *Street Rodder* magazine reader was planning to street rod a '38 Chevy coupe and he wanted to install a Chevy 261 cid six cylinder/1965 Powerglide transmission combination. He needed to know what flywheel, starter and torque converter would bring them together. Let me paraphrase at random from Patrick H. Dykes' proprietary flyer "Swap Your 216 Chevy 6 for a 235."

Using stock Chevy (six-cylinder) parts, any engine from 1937 through 1962 can be installed in any 1937–'62 passenger car or light truck chassis. "You will, however, need to pick the right parts." If you are installing the later engine and a manual transmission in a 1937–'40 passenger car, you will need a 1941–'54 wide-tooth (#139) 6-bolt flywheel off any 216, 235 or 261. Use the original bellhousing and starter. The Chevy 6-volt starter will

work even if you're converting the car to 12 volts. The fine tooth (#165) 1955 and later flywheel will not mesh with the original starter. Do not use it.

Complaints about Chevy starter alignment problems are common, but one of the least known causes involves the use of improper block attachment bolts. Genuine GM starter bolts have a knurled shank. (Knurled bolts better align the starter and prevent engine torque from disengaging the drive gear from the ring gear.) Furthermore, the OEM 3/8-16 fasteners are precisely measured—two are 4.656-inches long, and the shorter is 1.875-inches long. These specifications are not arbitrary. Fourteen-inch flywheels require starter noses with staggered mounting bolt patterns, but the smaller 12.75-inch flywheels use a nose with a straight-across bolt pattern. Both sizes were used on 327 and 350 CID engines.

Twelve-volt, high-torque starters for flatheads are available from Tri-C Engineering, 24927 Avenue Tibbitts, Unit F, Valencia, CA 91355, 661/295-1550.

DRIVE BELT TIPS

Drive belt damage or loss can be a minor disaster; especially if you're on the road without a spare! It is often caused by running up the engine rpm in the lower gears of a manual tranny. If this is one of your bad habits, and you don't seek counseling, at least look over these tips.

Belt slippage is usually a straightforward fix—it's loose, so apply a little "Belt Dressing" and tighten it. Or, it's slick, hard and worn out, so replace it. Sigh, street rodding life should be so simple! The truth is, there are several more insidious factors that are a reductive to adequate belt/pulley engagement.

One is a question of geometry—the belt must encircle at least 120 degrees of the perimeter of the pulley. Obviously, 180 degrees is fat city.

Not so common, but easily missed in the mix and match of V-belts and pulleys often found on street rod mills is that the belt is too wide for the corresponding pulley. Remember, it is the sides of a V-belt that bite the pulley. If the belt is too narrow, it will slip no matter how tightly you stretch it.

Another common cause is misalignment. Serpentine, multi-groove or V-groove, they must all lie in the same plane—a need easier stated than rectified given the propensity of street rodders to pirate their components from a variety of engine models and makes. Usually you can visually detect even a very slight misalignment with one eye closed, and then verify it by laying a straightedge across two (or more) pulleys.

Want a belt reduction on your small-block Chevy swap? A few years ago, *Street Rodder* magazine reader George Pareusan advised that his Chevy V-8 powered '47 Ford used alternator and air conditioner mounts that bolted to the front of the heads with bracketry from Bill's Hot Rod Co. (1064 E. Edna Pl., Covina, CA). He wanted two belts instead of three, so he used the 1957–'65 short water pump, the stock double pulleys, the vibration damper, and the power steering pump. "The secret is the double power steering pump pulley. I used the early pump with the bolt-on, keyed pulley. I went to Kragen Auto Parts and got a belt (PN 5815 or PN 15585). It is the rear belt that goes around the water pump, crank, alternator, and the rear groove of the power steering pump pulley. The front belt goes around the water pump, crank, and the air-conditioner compressor. If you have the large 350–400 cid vibration damper, you have to trim the lower bracket that bolts to the front of the block, as it was designed for the smaller 283–327 cid damper. Don't forget a top bracket for the pump. You can't mount the pump down low; try a mid-1970s Chevy V-8 Monza setup. That application mounts the pump on the driver's side cylinder head, almost level with the rocker cover."

EXHAUST TIPS (NO PUN INTENDED!)

The exhaust system unquestionably was the first thing to receive attention in the late 1940s when this hobby began to spread nationwide. The first modification of the Ford stocker the novice's dad allowed him to buy (and his mom discretely financed) was the installation of an extra pipe and a pair of Smitty mufflers. The first engine skill he learned (beyond brush painting the heads red and adding chrome acorn nut covers) was how to put pennies under the heat risers of the stock intake manifold.

Sixty years later, factory exhaust manifolds are still

seen on many street rods. Some are about as restrictive as you can get, and almost any switch to a free-flowing system will improve performance where it counts.

RAM'S HORNS

The popularity of the Chevy "Power Pak" exhaust manifolds has remained high. They are not nearly as bad as the "log" type, and the tight-fitting design admittedly makes most engine swaps easier. But there is still a price to pay in exhaust efficiency. The aftermarket has several designs, both cast and tubular, that outperform them significantly.

NOT TO BE OUTDONE

Ford has also offered cast iron performance exhaust manifolds and I saw a pricey pair at last year's L.A. Roadster Show. The show's patron base includes a high percentage of restorers, but street rod traditionalism shouldn't extend to the exhaust system when there are better alternatives on the market.

COOL IT

The design of some full-length store-bought headers for small-block Chevies places the collector too close to the starter, occasionally making a restart difficult during a pause on a road trip. A stainless steel sheet metal heat shield is easy to make, and will eliminate the problem. Or, buy a heat shield kit at your favorite speed shop.

PRUDENCE PAYS

Don't get carried away if you are going the custom-built header route. Most savvy exhaust system plumbers will tell you that tube diameter approximating the diameter of the exhaust valve, primary tube lengths under 40-inches and low restriction mufflers will deliver ample street performance.

THE TRADITIONAL FENTON

Costly custom-built headers for flatheads are very attractive, but those "Siamese" exhaust ports are difficult performance obstacles to overcome. The traditional street rodder should look at the cast-iron Fentons available from Patrick's and one of the

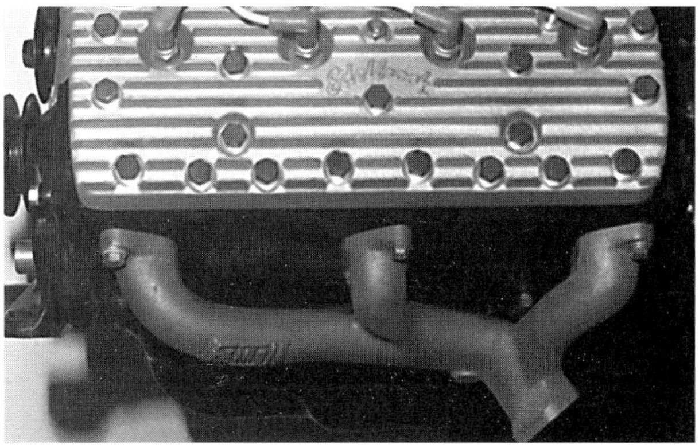

many varieties built by Red's Headers (22950 Bendar Lane, Ft. Bragg, CA 95437; 707-964-7733; www.reds-headers.com).

A BETTER FIT

You don't always get what you pay for, but that's not the case with premium-quality, extra-thick exhaust gaskets. To perfectly mate them to the predictably irregular surfaces of most header flanges, soak 'em thoroughly in water and slide 'em in while they're still juicy. Even the best gaskets take time to "set," however. Retighten the bolts a time or two in the days following the installation.

A GOOD ACT

Hedman Hedders has special locking header bolts called "Stage 8." They work.

CROSSING OVER

If you plan to go to the trouble and expense of custom headers for your street rod, don't forget the crossover tube. It should be of the same diameter as the exhaust pipes, and fitted in a foot or so from the ends of the collectors. The crossover not only adds a few horsepower, it softens an otherwise harsh tone. (I recognize the fact that the latter is in the ear of the beholder.)

STRESS ANALYSIS

Make sure those new mufflers you are hanging aren't placing an inappropriate stress on the headers. And don't buy the old doughnut shop rumor that tailpipes decrease any performance gains headers deliver.

LUBRICATION TIPS

Next to gasoline, oil is the most important fluid in your engine. A wise street rodder never takes it for granted.

Oil Viscosity—is one of those concepts that are sometimes only vaguely understood. Simply stated, viscosity refers to the ability of a liquid to flow. The higher the grading number assigned, the thicker the oil is, and the more slowly it flows. What's important to street rodders is that there are consequences to viscosity—the higher the numerical rating, the more horsepower it takes to pump the oil. That's a downside we can live with; street rods aren't often short on power. Some, though, have periodic bouts with overheating. The lower the numerical viscosity rating, the less efficient the oil is under high temperatures.

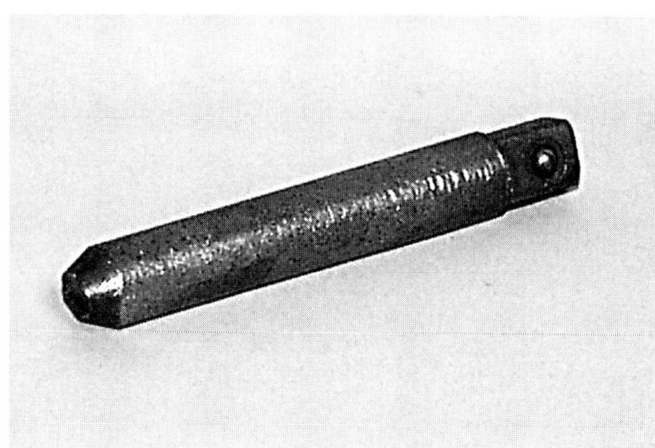

The Touch Test—Every time you drain engine or rearend oil, (or transmission fluid), let it settle for awhile, then run you finger tips along the bottom of the pan and feel for metal chips or shavings. Yes, it's a messy little chore, but real street rodders have inquiring minds. Seriously, think of it as part of your early warning system.

Prelubing—The best way to get oil up into the galleries of a fresh engine so that bearing surfaces are adequately lubricated, is to mechanically spin the oil pump driveshaft with a reversible, high torque drill motor until a pressure reading shows on an attached gauge. Doing so works out the air that's

trapped in the lubrication system before loads are placed on the components. Chevrolet dealers can order a "pre-oiling shaft" (PN 12368084) for such tasks.

For my Fords, I cut the female end off an old quarter-drive extension so that it can be chucked into a reversible drill motor. With a quarter-inch hex socket on a 12-inch quarter drive extension on the end of the oil pump driveshaft, I spin the pump for a few minutes. (Remember that the oil pump rotates counterclockwise, so be sure to switch the drill motor over.) I monitor the gauge while a buddy looks the engine over for oil leaks.

It's easy to forget that immediately after an oil change, the engine's oil galleries are likely to be empty. To prevent potential oil starvation upon fire-up, take the precaution of filling the new filter with oil just before you install it.

PCV Tips—The Positive Crankcase Ventilation (PCV) gadget is the basic emission control with which street rodders are confronted. It prevents crankcase blow-by from entering the atmosphere without choking off the crankcase. Essentially, it consists of a spring-loaded needle valve in a housing usually attached to the rocker cover with a hose to the intake manifold. When the engine is running, manifold vacuum unseats the needle valve and allows crankcase vapors to enter the plenum chamber directly. The vapors are then drawn into the combustion chamber along with the fuel/air mixture and burned. If there is an intake manifold backfire, the valve closes. If, however, the engine puts out more blow-by than the PCV can handle, the excess gases are routed to the air cleaner, combined with the fuel/air mixture, and eventually burned in the combustion chambers. When the engine is shut down, the spring holds the needle valve closed. The PCV has the least effect on high performance, and aside from cleaner air, significantly reduces oil fume residue in the engine compartment.

Flatheads, too, need a PCV. Drill and tap a hole in the manifold back where the old fuel pump was mounted to gain access to crankcase vapors. Install a screw-in PCV. Drill and tap a second hole between the carbs thus gaining access to the plenum chamber and install a fitting and connecting hose. (Always try to get in the center of any plenum. A location right on or near an individual intake will result in a lean mixture condition to that cylinder.) A filtered cap on the oil fill tube finishes the job.

Oil Pressure—A good rule of thumb for oil pressure in any street rod engine is a 10 psi increase for each 1000 rpm. A slight delay in getting a reading is not uncommon because the gauge is the last to show the pressurization, especially if you have a "garage queen" and you don't fire that puppy up very often. If you're not getting a reading of 25 or 30 psi after the engine temperature has normalized, it is possible there's a problem somewhere in the system.

Drip Proofing—Oil pan seepage is as frustrating for street rodders as it is for wives who despise a stained driveway. The stock plastic washers between the plug and the pan are adequate unless they are over-tightened and split. Aftermarket washers with an O-ring molded into them are available through auto parts stores and speed shops and promise a leak-free seal.

The rear main seal often gets blamed for oil leaks, but they can also appear if a cam plug or gallery plug isn't tight or if the oil pan isn't sealed against the block and rear main cap. It's just one more reason to pre-lube and build oil pressure in the engine while it's still on the stand.

Filtering—Want more filtering as well as a little extra oil in the lubricating system of that Chevy small-block without the hassle of extra plumbing? Pick up an overlength Fram (PH373) oil filter. Never buy an "economy" oil filter even though you plan to discard it long before an oil change is needed. Premium quality filters are worth a few dollars more.

Even a "garage queen" that doesn't rack up a significant annual mileage needs an oil and filter change every twelve months. Good maintenance requires an elimination of accumulated moisture in the engine.

Cooked Oil—You simply can't get away from it—street rods frequently have an overheating gremlin lurking somewhere. If that boil-over catches you on a hot summer day in traffic, change the oil at the first opportunity. Why? Motor oils cooked at boiling temperatures will in a relatively short time suffer a decrease in "oxidation resistance." Temperatures in the oil pan can be as much as 50 degrees above that showing on your coolant temp gauge. Motor oils begin to lose their oxidation resistance at 200 degrees. Fifty percent of the oxidation resistance is lost at 220 degrees.

Proper Startup—Avoid the impulse to over-rev a cold engine. It takes several seconds for the oil to thin out and course through the galleries to the bearings.

VIBRATION R & R

Engine-related vibrations plague even the tightest street rods. Isolating their cause is an exercise in patience at best, and a world of frustration at worst. Happily, a practicing street rodder can diagnose most of them himself. After you have identified exactly what is bugging you, you have to evaluate the problem under changing conditions. It is a process of elimination.

One of the major problems in vibration analysis is that the vibes migrate from the root source (the engine, the driveline, or the suspension) through the frame or the body and only then to your ears. By that time, however, they have been muffled by the interior trim, and an accurate identification of the root source is fuzzy at best.

First inspect the engine mounts. (So simple yet so often overlooked!) Next, check the belts—what may seem to be engine vibrations could well be belt damage or worn bearings in the alternator or A/C compressor. Unfortunately, the move to a single serpentine belt-drive makes this more difficult than it used to be.

If the source of the vibration is in the rotating and reciprocating components of the engine, clutch assembly, or torque converter, it will change perceptibly as you cycle through the gears. The next move is to find a safe, quiet area for a few test runs. Listen closely to the vibration or noise while you slowly accelerate; then coast along in neutral. Downshift, using engine compression for braking, then run the rpm up in neutral with your hand on the shifter.

Make a few abrupt stops, than a few gentle stops. Listen for vibrations or shudder while at rest with one foot on the brake. Slowly run the rpm up and then back off. Drive off again and steer gently left and right. If road and traffic conditions permit, weave a little. (Please be prudent when doing this!)

Make a few slow takeoffs and a few "jackrabbit starts" as they used to say. What do the vibrations feel like? What do they sound like? What do they respond to?

Steering, front suspension and braking vibrations typically transmit to the steering wheel. Rear end problems are usually felt in the set of your pants. Murphy's Law always rules!

Try to determine the frequency of the vibrations. Are they low or high? Is it a rumble, buzz or whistle? Every vibration has some particular characteristic that will help determine its cause and cure. That's why you have to keep duplicating it until it points you in the right direction.

TECHNIQUE: PRECISION METAL DISINTEGRATION

Sometimes late night work just doesn't pay off. Case in point: a pesky rocker arm cover leak. It is unbecoming for a street rodder and dangerous for a hot rod racer—oil can drip down on the header. A quick fix, right? Just change gaskets and tighten it up, right?

How many times has a simple task turned into a disaster? While tightening the quarter-twenty, a little too much torque elicited a dull snap. Yup. One broken bolt. A quick fix, right? Just reach for a center punch, the drill motor and the "Easy Outs." Sure. You know what happened next. First the drill wouldn't bite, then when it did; it skewed off

1. This is a close-up of a rocker cover bolt hole in the modified Boss 302 racing head I damaged. It might not look like much of a problem, but a missing bolt just above the header is a potential fire hazard.

2. The disintegrating head of the EDM machine can be rotated in three full 360 degree movements for compound angular positioning of the moly electrode. During the process, the electrode vibrates 3600 times a minute. Not only broken bolts, but taps, drill bits, reamers, hardened pins, bearings, set screws, heat treated studs, etc., can be re-moved easily and without damage to the workpiece or distortion of internal threads.

3. All metals that conduct electricity can be "disintegrated" by this process. Holes can be burned to within 0.002-inch tolerance by an experienced operator.

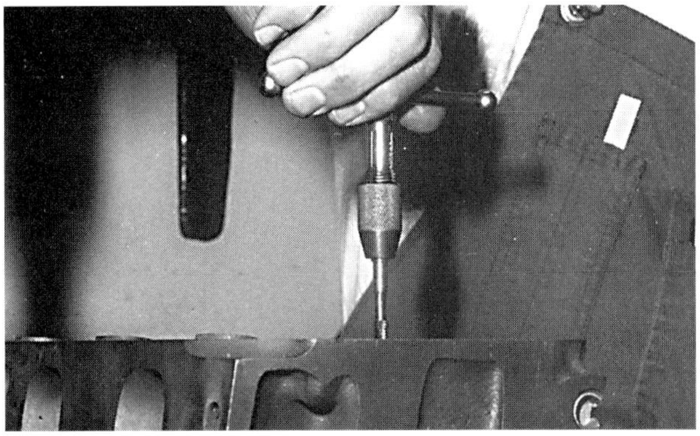

4. When the burn is done, the part of the broken bolt that didn't get "powderized" is picked out, and the internal threads chased with a tap.

sides and buggered up everything. Man, what a drag.

The next day I was down at my favorite machine shop for a professional fix to this "simple" problem. After a few minutes conversation, I knew what Alan Hail had in mind—precision electric discharge machining (EDM). I was on my way.

Tom Grafton, proprietor of Jerry's Drill and Tap Removal in Santa Ana, California gave me a brief introduction to the EDM. Tom said it was developed in the early 1950s and removing a broken bolt or a tap only scratches the surface of the technique. "We can drill square holes!"

He explained that the cutting action of metal disintegration is accomplished by creating a series of intermittent electric arcs that break down the hardest metals. The electrode (which functions as the cutting tool) is held in the head of the disintegrator and vibrates as it cuts. During the process, a coolant pumped through the electrode washes away the metal, which has been reduced to powder.

Jerry led me to one of the machines that vaguely resembled a mill. The most noticeable difference is that the electrode hanging out of the chuck is a piece of molybdenum tubing.

"That's what we'll burn the broken bolt out with. What we do is burn out everything that's in the original hole—bolt, tap or whatever, within a few thousandths of the minor thread diameter leaving only the coil of the thread from the broken bolt. Then we go in there with a small chisel—sharpened at just the right angle—and start picking the pieces out. Once that is done, we run a tap through the original hole and chase its threads."

He continued speaking as he placed the cylinder head on the machine table (which I later learned is Blanchard ground for exact perpendicular mounting of the superstructure and will support a workload of 8,000 pounds). "Anything that conducts electricity can be salvaged in this way. We've even burned broken drill bits out of graphite. Graphite is extra heat resistant so we have to kick the amps way up. We can change the size of the electrode if more heat is needed. We can put holes in parts as small as 0.020-inch or as large as an inch and a half!"

As he turned the machine on and advised me to back away, it looked ever so much like a combination drill press and arc welder with sparks flying all about. As I was to learn, the mobility built into the vertical column, horizontal ram and head housing provides adaptability that makes the work capabilities almost limitless. "It's very versatile. Anything we can put on the table gets done that way, but if it's too tall, we'll put it on the floor next to the machine. Some industrial jobs are so heavy we can't get them off the delivery truck, so we bring the machine over to the door and run it there. Anything that is accessible to the machine, we can do. The head is adjustable 360 degrees and we have auxiliary grounds. We could roll the machine next to your car and do this work on the engine without ever removing it, but that would be expensive, because moving the machine would take more than an hour's labor."

The molybdenum electrodes range from .020 to 1.25-inch, and are available round, square or hexagonal. (A square electrode disintegrates a square hole, etc., etc.) And talk about fast—a broken 1/2-13 tap one inch deep can be removed in less than a minute!

All tool and die steels, titanium, Alnico, stainless steel, stellite, cobalt, Inconel, Hastelloy, copper and aluminum can be cut. And those of you into modern ultra-machined street rods could order a dovetail put into a hardened steel widget without annealing it. But as you can see, my needs were much less exotic. And in less time than it took you to read this far, my repaired head was back in the bed of my truck.

I like this kind of stuff, and I've always assumed

1. Kit instructions specify the required drill size; use a sharp bit at slow speed. Insure your alignment with a pair of bolts to sight in the proper drilling angle. Keep the chips out of everything.

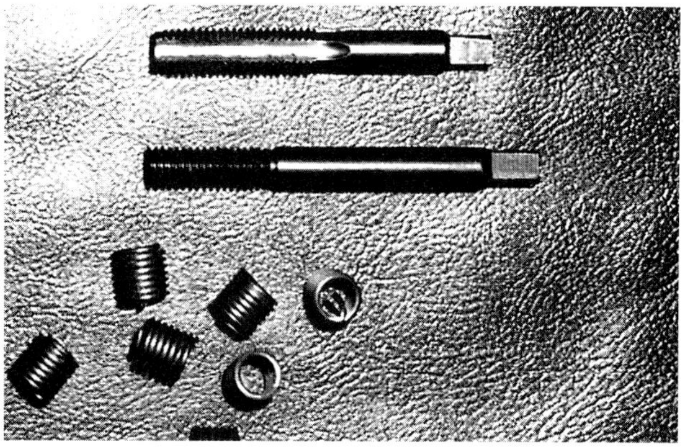

2. Heli-Coils use a special tap that is included in the kit. The alignment angle is critical.

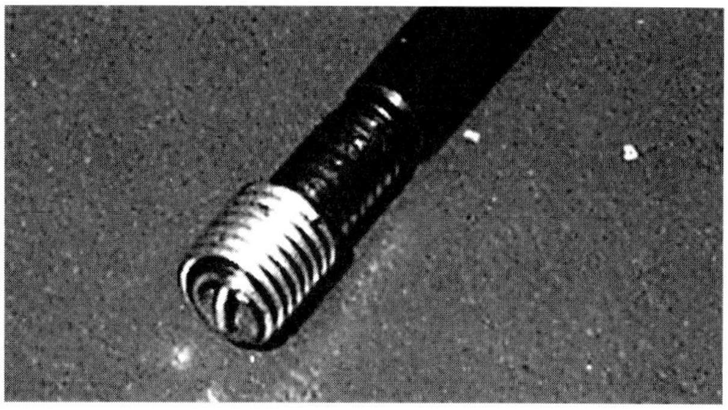

3. The Heli-Coil is fitted on the inserting tool . . .

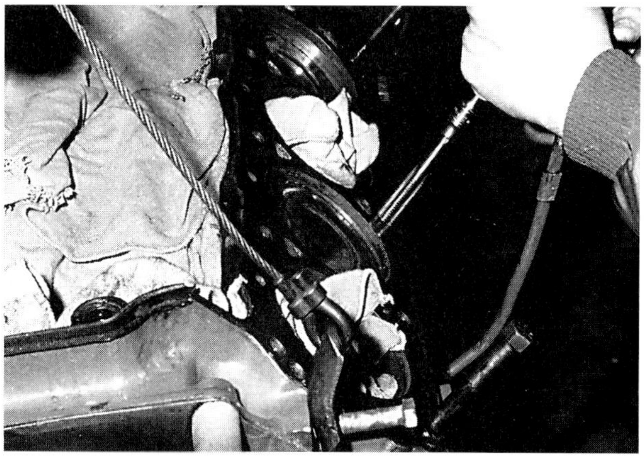

. . . . and screwed all the way down. Be sure the coil clears the top of the surface, but not more than a half thread deep.

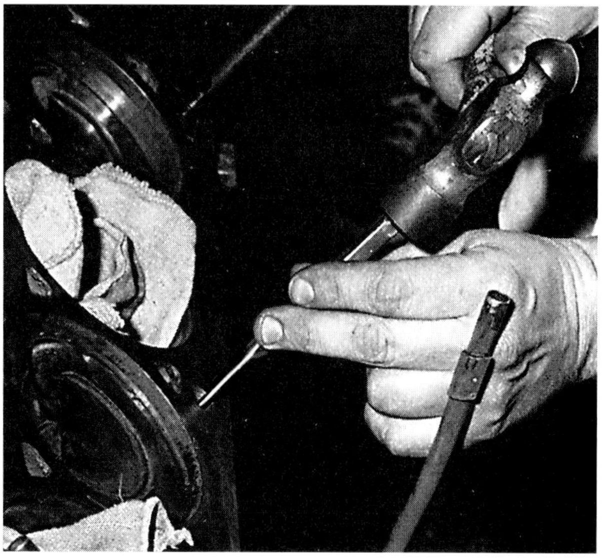

4. Break off the tang and the job is completed.

most of you do, too. You don't have to stretch your imagination to see its applicability to street rod building and maintenance. But for now, just tuck the info away in your random access memory. This service is usually available in any large metropolitan area. When you need it, check in the Yellow Pages under Taps—Extracting or Machine Shops. This isn't something you only read about and wish you had access to. You're gonna need it someday.

TECHNIQUE: THREAD REPAIR

Fortunately for heavy-handed street rodders who ruin the threads on some critical part, there's an inexpensive product that will often bail them out of trouble. Precision thread helical inserts were designed to repair stripped or damaged threads in any metal and even plastic. They are preferred over a tap because the original screw hole diameter is maintained as opposed to tapping a larger diameter hole with the same pitch as the original thread. The best known of the thread repair kits is manufactured by Emhart Products, a division of Black & Decker, Inc. and is known as Heli-Coil. These helical inserts are a coil of diamond-shaped 18-8 stainless steel wire, the inside of which provides the mating threads for a bolt. The coils themselves are quickly and easily sunk into the thread-damaged bolt hole in whatever you messed up. Kits can be purchased for specific size repairs and are available in National Coarse, National Fine, Metric and spark plug threads.

TECHNIQUES: GOOD BALANCE AND TIMING IS EVERYTHING

Changes, changes, changes. Yeah I know, "The times they are a-changing." Changes seem to be a common occurrence with guys who have spent a lot of time with the same street rod. Some are to be expected—mechanical components in motion do wear out. But then there are those other changes. New and seemingly better components come forth from both Detroit and the aftermarket street rod industry. Add to those the changes in trends. All of them are "subjects of interest" and some of them eventually tempt street rodders to continue the unending pursuit of as much mechanical perfection as possible.

I was perfectly happy with the Ford top-loader 4-speed transmission that was bolted to the small-block Ford 302 that powered my coupe at the time. Nevertheless, I was knee-deep in the construction of another project, a 1941 Ford pickup. The 4-speed appealed to my sense of "what belongs" in a working pickup.

The downside was that the engine in the coupe had been precision balanced with the flywheel and clutch assembly. Now, factory delivered small-block Fords are externally balanced, that is, everything connected to the crankshaft must be included in the balancing operation. I was in a quandary, especially since I had planned to replace the existing harmonic balancer for a thinner one that provided additional radiator clearance for the special water pump and timing chain cover scheduled for installation. (Experienced street rodders well know how one change invariably leads to subsequent changes!) I was afraid I would have to blow the engine apart for a rebalance.

Before rolling out my engine stand, however, I decided to talk to the Alan Hail, an automotive machinist and my longtime mentor. He opined that a complete rebalance was not necessarily required. Perhaps the new (thinner) harmonic balancer and

the torque converter flex plate (which was much lighter than the flywheel and clutch assembly) could be balanced and weight-matched to the existing components.

My next conversation was with Andy Miron, the weigh master who had previously handled conventional balancing for me. Andy said he could spin the old harmonic balancer, flywheel and clutch assembly and take dynamic readings. He would than spin the new balancer and flex plate for another set of readings. Finally, and with a little drilling and weight additions here and there, he could match the dynamic balance of the new components to the old components. He was confident he could get close enough to keep my engine happy. In fact, Andy boasted that he had never had any complaints on a rebalance match. The cost was minimal and the time and effort saved was maximal, so I'm passing the information on for those of you who someday find yourself in a similar predicament.

Eventually the engine and new automatic

transmission was back in the coupe and was ready for re-timing. Whoa! With all the accessories (AC compressor, power steering pump and alternator) plus new bracketry and the serpentine belt needed to drive everything, I couldn't get a straight shot at the indexing marks on the harmonic balancer! Besides, the new timing chain cover did not have a timing pointer. I would have to re-index the balancer and make a new pointer.

Think that encompasses some tricky machine work? Naw. I just dug out my old degree wheel and homemade "Positive Stop" plug to relocate the Top Dead Center (TDC) mark so that I could correctly install a timing tape anywhere I needed it on the balancer. Now, you may not have a degree wheel, or you may never have even heard of the Positive Stop Method, but I'll bet (if you are a hands-on street rodder) that you are going to need both someday. As street rods grow ever more luxurious, and more and more accouterments are deemed "necessary" chances are very good that you will need a new window of access to poke your timing light through for accurate ignition timing. Here's what you do . . .

First take an old spark plug and knock off the porcelain and gut the electrode pocket. Then braze a piece of steel rod in the pocket. That done remove all the spark plugs from the engine and rotate No. 1 piston up to the top of the stroke. (Turn the engine over by hand, using a ratchet on the bolt that secures the balancer to the crank.) Measure the distance from the top of No. 1 piston to the bottom of the plug thread and add a quarter inch. This is how long the rod brazed into the indexing spark plug should be. Trim it to length, round it off and smooth the end because the piston will come into contact with the plug when you rotate the engine. Shazam! You now have your very own Positive Stop tool. Run it into the No. 1 cylinder sparkplug hole.

Next, make up a steel strap that can be bolted across the diameter of the balancer. Drill a 3/8-inch hole in the center of the strap that lines up with the center of the balancer and square it off so that you can insert the drive lug of a long handle ratchet into it.

Then, locate the best spot on your harmonic balancer for easy, straight-on reading with your timing light and make up a timing pointer. You can use welding rod or a piece of aluminum--anything that can be attached to the timing chain cover and will be a permanent reference point.

Now rotate the crankshaft slowly in a clockwise direction until the piston lightly touches the rod in the indexing plug. Center punch a dot on the balancer in line with your new pointer. Then rotate the crankshaft in a counterclockwise direction until the piston again touches the indexing plug rod. Using your reference point on the timing chain cover, center punch a second dot on the balancer.

Note the distance between the two marks in degrees if you have a degree wheel, or in fractions of an inch if you use a narrow, flexible machinist's tape measure. TDC will be exactly half way between the two readings. That is, if the distance between the two stops of the balancer was 70 degrees, TDC would be 35 degrees from either stop. If the distance was two inches, TDC would be one inch

from either stop.

Recheck your findings, and when you're confident you have located Top Dead Center where you can get a straight shot at it with your timing light, double punch the center point and stick on the timing tape with the zero over the punch marks.

This brings us to one final consideration—the timing light itself. Now you can walk into any discount auto parts house and buy a perfectly good timing light at a very reasonable price. Most are simple enough to use. You hook up the leads, aim the light at the engine's calibrated indicator and timing pointer, and with the engine idling at normal operating temperature, ascertain that the marks on the balancer and the timing pointer are aligned. That's about it for the weekend cruiser, but not for the performance-oriented street rodder. He also wants to know (or to establish) the full advance of his ignition timing at some predetermined point, usually about 3000 rpm. He cranks the engine up to that rpm and reads the timing directly from the indicator or timing tape as the strobe flashes.

Technology moves on however. Some years ago I was watching Chris McDivitt, a high-performance mechanic, tuning a customer's car with a big red

Snap-On inductive timing light, and he was twisting dials. Hmmm, I thought, my timing light ain't got no stinkin' dials! Needless to say, I have had one of my own ever since. I'm sure there are even better models nowadays (at least the model number has changed) but mine hasn't once failed me in my home garage, the dusty dry lakes or the Salt Flats.

Now I don't want to rewrite the instruction manual that comes with the light. But what isn't in the manual is McDivitt's commentary as he showed me how the light functioned with a centrifugal advance, electronic ignition such as the Hays/Stinger I've used in the coupe for decades.

"Measure the base timing first. Clean and highlight the timing mark on the balancer at TDC and the pointer. Use silver automotive touch-up paint, the kind that comes in a little brush-cap bottle. It dries fast and reflects the timing strobe well. Run the engine at idle and aim the timing light at the pointers. Turn the adjustment knob on the light until the marks are aligned. Read the number from the adjustment knob; this is the base timing (in degrees). Record this number for use later."

To measure the centrifugal advance, "Aim the timing light at the marks while running the engine at approximately 3000 rpm. Turn the adjustment knob until the marks are aligned. Momentarily increase engine speed to see if the timing marks move any more. If they do, readjust the timing light knob to align the marks. Note the rpm at which the timing marks stop moving and note the number indicated on the timing light. To calculate the centrifugal advance alone, subtract the measured base timing from the measured advance. To calculate the total advance, add the base timing to the measured advance."

This is the kind of stuff that hands-on rodders like to know, and this is the tool to acquire that knowledge. Precision balancing and precision ignition timing makes the street rod tango a little more fun. My engine sounded as smooth as ever with no perceptible vibration, and I swear on a stack of Motor Manuals, it has a few more beans.

A PARTING THOUGHT ABOUT PERFORMANCE

One horsepower is a mere pittance in terms of motor muscle to hear folks talk, but not if you go by the definition of Scotsman James Watt (inventor of the steam engine). One horsepower is the ability to lift 33,000 pounds one foot off the ground in one minute!

Chapter 2

TRANSMISSION TIPS

AUTOMATIC TRANSMISSIONS

For more than thirty years the automatic has been the transmission of choice for the street rodder who values cruising comfort, to say nothing of the desire to keep his wife (and co-pilot) interested in participating in rod runs with drive times longer than sixty minutes. The automatic, in its modern, high tech state with electronic sensors and overdrive, is undoubtedly more efficient, but the other side of the coin is that it is more complicated than the automatics of the late 1960s through the mid-1980s. Where many performance modifications are out of the skill range of the average street rodder, there are still a number of timeless considerations.

KEEP IT COOL

The optimum temperature for automatic transmission fluid is 180 degrees, and although a brief rise to 233 degrees is acceptable, sustained operation above that can drastically shorten transmission life. A standalone aftermarket trans cooler is a wise investment. Just don't mount your new trans cooler too close to any part of the exhaust system. A six-inch air gap is the minimum.

BUT NOT TOO COOL

On the other hand, if the temperature within a transmission built with stock clearances is below 180 degrees, the fluid may not thin down enough to provide adequate lubrication. Although this is not a problem a street rodder would often incur, it is at least something of which he should be aware if he lives in very cold country.

DRAIN PAIN

Several makes and models of popular automatic trannies do not incorporate a fluid drain-out bung in the pan. Avoid most of the mess; install a universal drain-plug kit when next you drop the pan. It is inexpensive and available in most auto parts stores.

TAP DANCE

Full-time manifold vacuum is required for correct operation of an automatic transmission's vacuum modulator. If your new carb doesn't have a vacuum hose port on the base plate, drill and tap the manifold for a fitting.

STALL SPEED

The automatic transmission torque converter's stall speed is the maximum rpm reached before the driving wheels are engaged. It is somewhere between 1200 and 1500 rpm with a stock converter, but a street rodder with a yen for a bit of high performance, can easily go overboard when choosing a replacement. An aftermarket unit in the neighborhood of 500 to 800 more rpm than stock can safely scratch that itch.

FOMOCO UPGRADE

The Ford automatic overdrive transmission (AOD) will bolt to any six-bolt bellhousing 289/302/351W cid small-block that is compatible with the older C-4 automatic, but there are some things to remember. It is the same length as the C-4 but with a different mount location. The AOD will accept the 28-spline C-4 yoke and driveshaft. But

there's more . . .The same engine spacer plate and starter can be used, just make sure the flex-plate matches the torque converter. (Street rodders invariably have a parts surplus, and they aren't always properly labeled.)

COUNT THE TEETH

Pre-1980 Ford small-block automatic transmissions used flex-plates with 157 tooth ring gears and 28.2-ounce balancing weights, (E0AZ-6375-A). However, there was a changeover in 1980–81, and subsequent production motors use a flex-plate (pictured) with a 164-tooth ring gear and 50-ounce balancing weights, (E2AZ-6375-A). Total Performance (44020 N. Grosbeck Hwy. Mt. Clemens, MI 48043, (313) 468-FORD) catalogs an early 157-tooth flexplate with 50-ounce weight (PN TP-6375-83BR) when a hybrid is needed.

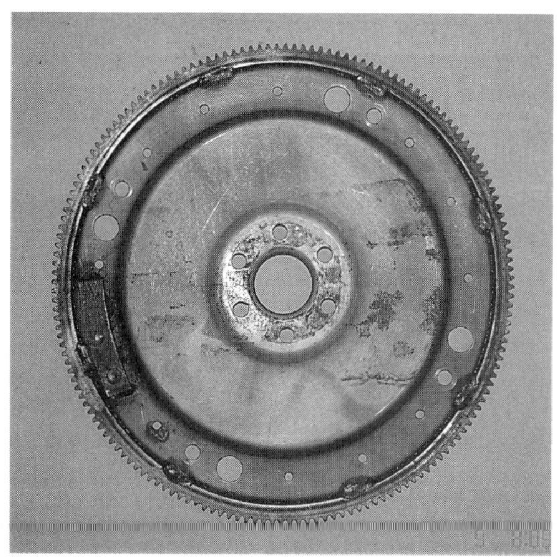

TV TIME

Most Ford AOD transmissions were originally mated to fuel-injected engines. When they are hooked up to a carbureted engine in a street rod, the throttle-valve (TV) control cable that

determines shift points won't work properly. If you can find the mechanical linkage and bracket from a 1979–'80 LTD, you can safely use it with your carb, but in lieu of that, head for the aftermarket! Just remember, there are specific linkage adjustments that are critical to the AOD. If you want to make them yourself, they are covered at length in the factory shop manuals available through the local Ford dealership.

GEARED TO GO

The Ford AOD has a final gear ratio of 0.67:1 as opposed to the 1:1 gear in a non-overdrive transmission. It also has what is known as a "split-torque" gear train, one that reduces the hydraulic losses associated with a torque converter. That is, when in overdrive, all torque is transmitted through a direct drive input shaft bypassing the converter.

Nevertheless, a street rod with a tall final drive ratio, plus big rear tires may "lag" a little before the manual kickdown takes over. This is normal, but there can be an added wear factor. For that reason, a rear gear ratio in the 3.50:1 range is prudent. If more acceleration is desired, get a little closer to the 4.00:1 range.

FORD TORQUE CONVERTERS

The Mustang 5.00 L HO engine torque converters have long been considered the best stockers for performance street rod use. If availability turns out to be a problem, jot down the following address in your journal, TCI Automotive, One TCI Dr., Ashland, MS 38603, 601/224-8972. TCI is a great resource for performance AOD transmissions, torque converters and accessories.

GM's TH 200R4

If you are planning to run a Chevy small-block, and the chassis is a bit tight (as in the Model A years), certainly consider the TH 200R4. It is significantly thinner in girth than its big (and more popular) brother, the TH 700R4. If high performance is anticipated, however, it should be reworked by a competent transmission technician who can install the proper valve body and other heavy-duty components required for strength.

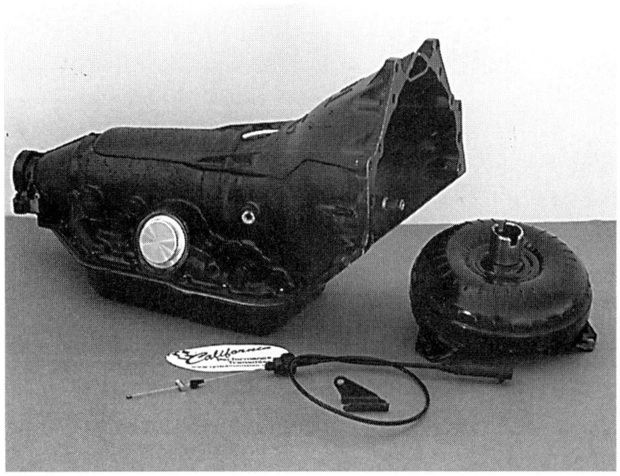

The Turbo 200 automatic transmission version that comes mated to many street rod compatible GM V-6 engines is worth keeping if you're a laid-back street cruiser. However, make sure the wear-prone aluminum governor cover on some models is replaced with the optional plastic piece. This, again, is a job for the automatic transmission shop.

GM's TH-350

The 350 Turbo Hydra-Matic automatic trans has been a long-time street rodding favorite. In fact, although there are no definitive statistics, I'll still wager it sits in the middle of more street rod frames than any other transmission, automatic or manual. If you are running a healthy one, there is no economic or performance reason to swap it out. On the other hand. . .

GM's TH 700R4

. . . these days of astronomic fuel prices may lead the builder of a fresh street rod to consider the more fuel efficient TH 700R4 with a converter lockup control. Popular street rodding choices have been the TH 700R4 that was originally found in the 1992–'93 Corvettes and Camaro Trans-Ams with the round five-prong plug on the driver's side. If you are replacing a Powerglide with a 700R4, although the stock front yoke on the driveshaft will fit, the driveshaft itself will have to be shortened and rebalanced.

Although most small-block Chevy flexplates (pictured at left) are straightforward (unlike Ford plates), recommended upgrades for the 700R4 include a small-block flexplate with a dual bolt pattern for use with older engines. A deeper fluid pan and an aftermarket shift kit are also worth considering.

CALL THE BUFFALO

Die-hard Chevy six-banger enthusiasts often have problems finding drivetrain adaptation hardware. If you are one of them, jot this down: Buffalo Enterprises, 25625 Dahl road, Arlington, WA 98223 (360) 652-7684. These folks have one-piece, Blanchard-ground, steel adapters for the Chevy 235/261 and GMC 228/302 engines to Powerglide, TH350, and TH400 automatic transmissions, as well as scatter shields and custom flexplates for the GMC and Chevy six-cylinder.

NAIL IT

When it comes to power plants from another era, a few engines other than the flathead V-8s and the GM six-bangers still have a following among rod builders. To whit, the nail head Buick V-8s of the early 1960s. However, the two-speed Dyna-Flow automatic transmission originally mated to them just hasn't retained the same traditional mystic. Fortunately, the 1965–'66 Turbo series will bolt right up.

MOPAR ADAPTATION

In the "Anything Can Be Mated Together" Department is the adaptation of a MoPar 727 Torqueflite from the A series engines (273/318/360 cid) to the short-bellhousing 331/354/392 cid Hemi. It's not exactly a piece of cake, but if you are willing and capable of doing a bit of work yourself, drop a line to early Hemi guru Gary Stauffer, 1150 Ryan Ct. S.W., West Linn, Oregon 97068. He offers a kit and instructions.

DROP ANCHOR

When it's time to torque the flex-plate (or flywheel, for that matter) bolts to the crank flange; anchor it to the block with a few inches of lightweight chain. If you don't have an impact wrench that directly centers the force, do the same for removal of such fasteners.

BEAN COUNTING

Don't be penny-wise and pound-foolish with regard to kickdown linkage. It isn't necessary to spend a Saturday looking for and pulling stock components at the local auto dismantler, no matter how penurious you are. The street rod components industry offers a variety of universal throttle linkage/kickdown cable kits at reasonable prices, plus instructions.

MANUAL TRANSMISSIONS

There is a relatively small, but adamant percentage of street rodders who still embrace the manual transmission and clutch assembly. Installation and maintenance concerns are just as prominent, however, be they novice or veteran.

CHECK 'EM OUT

Inasmuch as there are a variety of clutch cover heights and disc thicknesses, determining the correct throwout bearing for your clutch/transmission marriage is important to ensure proper fit and maximum leverage. If you've installed the right bearing, the outside end of the clutch arm will point slightly forward when

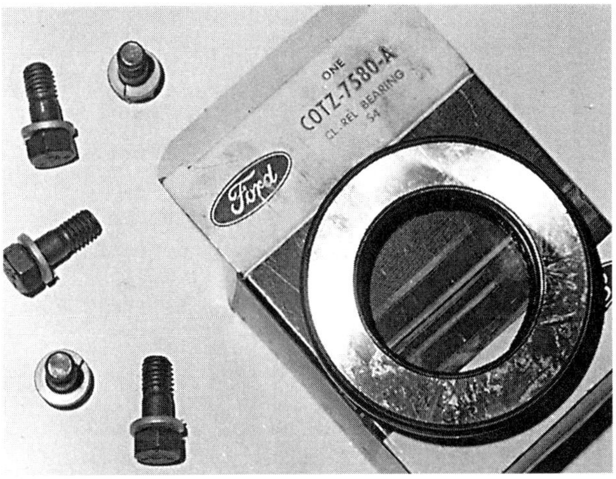

assembled. New clutch bolts are always a wise move.

A PRECISE INTERFACE

If you don't know why you are breaking transmission parts, get the engine block or bellhousing resurfaced and trued before you reinstall the repaired transmission. Unlike the attachment of an automatic transmission, cylinder block interface is critical with a manual transmission bellhousing in a high-performance setting. (The automatic trans flexplate compensates for manufacturing tolerances because it actually does flex slightly when it is rotating.)

EASY IN

Even with the help of a buddy, installing a transmission is almost like threading a needle. It's a darn sight easier if you make up a pair of homemade alignment studs. Scrounge up two three-inch long Grade 3 (soft) bolts that match the threads in the cylinder block. Cut the heads off and hacksaw a screwdriver-sized slot in the decapitated ends. Use your bench grinder to remove the sharp edges.

EXTEND THE SERVICE LIFE

Always keep your clutch properly adjusted. You should have from a half to a full inch freeplay with the pedal in the engaged position.

FORD MANUAL TRANSMISSIONS

A late 1960s Mustang replacement clutch, flywheel and bellhousing (the casting number for the latter is E6ZR6394AA) will mate a 1985 or later T5 manual transmission to a small-block Ford 302. Kits from Pony Cars, 1906 Quaker Ridge Pl., Ontario, CA 91761 (909/923-2357) make the job a lot easier.

A REAL FLOOR SHIFTER

A 1976–'77 Jeep CJ-5 transmission top-mounted gear selector will function on the exceptionally strong Ford 4-speed "Top Loader" transmissions used in late 1960s–early 1970s high-performance Ford products of that era. The case only has to be mildly clearanced after removal of the side-shifting forks.

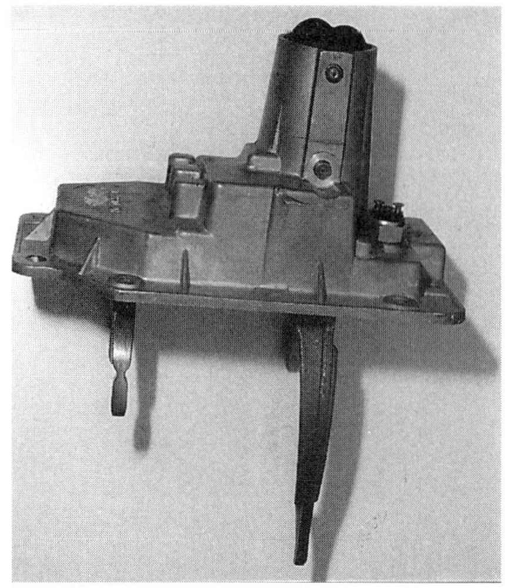

Another interesting "late model" Ford manual transmission is the one that premiered in 1981 Ford trucks. It is a 3-speed with overdrive, and is more than appropriate-looking for a flathead-powered pre-WWII Ford pickup, and the gearing is better than most for flathead power. A Jeep top shifter also fits the box with minor grinding, and looks much like the original flathead equipment.

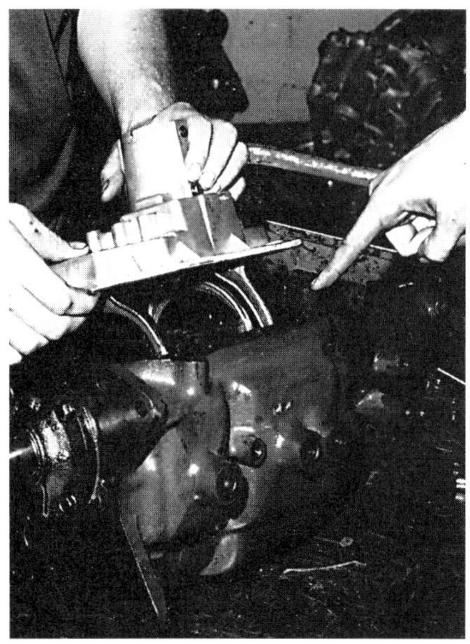

The overdrive works out to about 29 percent and depending on what rear gear ratio and tires are used, a final drive ratio should be compatible with the torque output of a mild flathead. There are two different gear sets available, but the most preferable has a 3.00:1 first; a 1.92:1 second, and a 1.1:1 third. This setup works well with the 1966–'77 Bronco rear which often comes with either a 4.11 or 3.50:1 rear gear.

FLATHEAD MARRIAGES

Several 1949–'53 Ford and Mercury bellhousings can be used on an 8BA block to mate the super-strong 1970s-era Ford Top Loader gearbox if it has the dual bolt pattern.

Either the rare late 1951 Mercury 360-degree cast iron bellhousing (top photo, next page) or the stamped-steel housing with matching starter support plate that came behind passenger car and truck engines will work. Since there are varieties, buy them as a unit. Fortunately, they are still available along the swap meet trail and vintage parts emporiums. Typically, the front bearing retainer has to be turned down to fit.

If it's one of the late-model T-5 manual

transmissions that turn you on, check with Cornhusker Rod & Custom (RR 1 Box 47, Alexandria, NE 68303; http://roadsters.com/cornhusker) for their series of bellhousing adapter plates.

SPEAKING OF WHICH

Do not neglect to check the threads in an aluminum late-model bellhousing. They are often damaged and need a Helicoil fix.

FORGET IT!

Even if you have one, don't even consider using the stock flathead flywheel. Most any of the high quality aftermarket flywheels can be used as the starting point when you are ready to build a compatible clutch assembly. Aftermarket flywheels for flatheads are drilled to accept a variety of pressure plates; just make sure the 10- or 11-inch pressure plate pattern you pick matches the flywheel you plan to use, and of course the hub in the clutch disc has to match your transmission input shaft.

SLAVISH DEVOTION

A hydraulic clutch master cylinder and slave unit from a GM or Dodge pickup, 1962–'69 vintage, and a bit of bracket R&D will take care of the external operation.

WEIGHT REDUCTION

Got a heavy clutch pedal? Lengthen the arm that is actuated by the link from the pedal. By the way, driving technique has an effect on clutch life. For instance, a great deal of wear on the clutch can be saved by simply downshifting less. (I think some street rod pilots secretly want to be Le Mans drivers.)

DOUBLE YOUR PLEASURE

The ideal way to slow a car with a manual transmission is to use the brakes and then double clutch directly to the desired gear, without going through the intermediate gears. Double clutching reduces transmission and driveline wear, but gear speeds in the transmission must be matched to engine rpm at the instant the clutch is released so that slippage is minimized. The technique is quickly learned with a little practice. Your shift should be timed so you release the clutch at the exact moment you want to get back on the throttle.

WEAR-OUT FACTORS

Other causes of rapid clutch wear are a warped flywheel, pressure plate or disc; a leaking rear main seal, improper adjustment; and driving with one foot resting on the clutch pedal.

DON'T FORGET

When the original 1940–'48 Ford pedal assembly designed for a column shift won't operate the clutch linkage for a modern manual trans swap, check out the 1937–'39 assembly.

MORE TRANNY CHOICES

If you're thinking of using a three-speed manual gearbox behind Ford power, there are two models of 1970s era Ford-design transmissions that could be considered. The preferred one for high performance street use has a metal tag bolted to the gear case stamped "RAT." It was generally used behind vehicles equipped with a 390 cid engine. It has the following "street-friendly" gear ratios: 2.42:1 first gear, 1.61:1 second gear, and of course a 1:1 (direct) third gear.

CHECK IT OUT FIRST

Pull the inspection cover and do a visual inspection of the gear and synchronizer teeth before purchase. Chipped or blue gear teeth are a no-no. And, as with any recycled complex automotive unit, it is always wise to verify return privileges in writing, and have the box professionally inspected before installation, not after!

NOT THE PERSIAN BRAND

The Ford manual 3-speed with overdrive (called RUG) looks very similar to the regular 4-speed top loader. To differentiate, mark the input and output shafts and count the rotations. The output shaft of the overdrive unit rotates about 1.25 times that of the input shaft. The 4-speed will only turn one revolution. When the Jeep gear selector is installed on the Ford RUG (3-speed with overdrive) you get this H shift pattern:

1 OD
Neutral
2 3

THE BRASS RINGS

Synchronizers are devices in manual transmissions that assure gears about to be meshed (as in going from first to second) are revolving at synchronized speeds thus avoiding the sickening sound of grinding. The synchronizer rings are made of brass, and as such they are (by design) "consumable," that is, they are expected to wear out. Many of the older muscle car transmissions are still popular with street rodders, but such parts are hard to find. If a Ford Top Loader is to be rebuilt, surely some parts will be needed. Contact Dan Williams Toploader Transmissions, 206 E. Dogwood Dr., Franklin, NC 28734. Dan's website is www.toploader transmissions.com.

WANNA GO LATE MODEL?

The early 1990s Mustang Borg Warner T5OD manual transmission is popular with street rodders. The clutch release is operated hydraulically on some models, while others use a mechanical system in which a clutch cable actuates the clutch release lever and throw-out bearing. The hydraulic system uses a slave cylinder to actuate the clutch throw-out bearing. The throw-out bearing is designed to consistently run against the diaphragm, that is, just loose contact with the clutch diaphragm. If there isn't enough free play at the pedal, space the slave cylinder back.

In a stock hydraulic system (after verifying an adequate fluid level) run the engine at idle speed with the transmission in neutral. Then dip the

clutch pedal, wait nine seconds, and shift into reverse. Everything should be quiet, that is, no grinding noises. Any such sounds indicate a problem with the pressure plate or disc.

To check for complete clutch release, hold the clutch pedal about a half-inch from the floor and shift between first and reverse several times. The shift should be smooth. Slave cylinder pushrod travel should be as least 15.5 mm. Always ensure complete clutch release before you leave home.

I'VE SAID IT BEFORE

There are two ways to bring a modern engine and manual transmission together—stock parts or aftermarket high-performance parts. The prudent rod builder, particularly the novice, is best advised to look among the factory-mated engine and transmission packages on display at the automotive dismantlers. Everything in between the two is compatible—bellhousing, flywheel, pressure plate, clutch disc, throughout bearing and sometimes, even clutch linkage. Everything fits together perfectly. On the other hand, the adventuresome like to mix it up. If you have (or have a hankering for) an engine and transmission that have never met before, drop a line to McLeod Industries, 1600 Sierra Madre Circle, Placentia, CA 92870 for their latest catalog. It lists list flywheel and disc/pressure compatibility within company products.

A BETTER FIT

SAC Hot Rod Products (625-35 West Katella Ave., Orange, CA 92867-4619; 714/997-3433; www.sachotrod.com) has a "chassis widener," a compact pre-formed piece that is welded into the existing frame and allows working space for transmissions (automatic and manual) that have side shift linkage that conflicts with the chassis X-member.

Chapter 3

REAREND TIPS & TECHNIQUES

FORD REARENDS

There have been a number of Ford-designed rearends since the old torque tube type finally went out of production in the late 1940s. Not all of the "late models" are good candidates for street rod use, but a well-versed rodding practitioner should be aware of them. It is the removable differential gear type introduced in the late 1950s that has, without a doubt, elbowed its way to the front of the line. The most common and popular of which are the 8- and 9-inch ring gear types with 28-spline spline axles, and available in a wide variety of axle lengths.

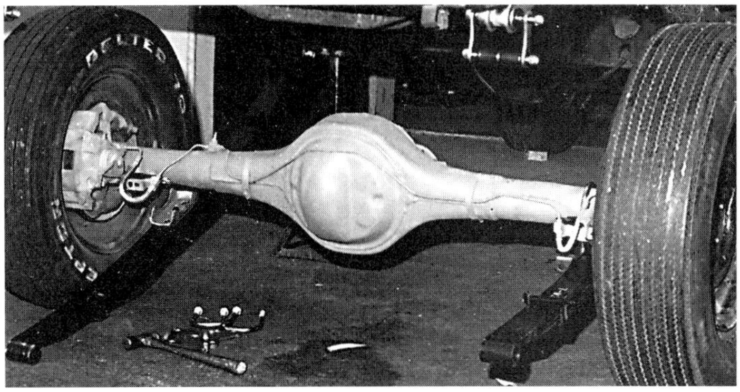

The first task to master is that of determining which is an 8-inch, and which is a 9-inch. The 9-inch is pictured, and it is easily recognized by its girth. Once you see the two together you can store the images in your random access memory. In the meantime, to be absolutely certain, just remember the bottom two bolts on the third member of an 8-inch gear carrier can be removed with a ratchet, extension and socket. If you look closely at the business side of a 9-inch, you'll see that only an end wrench can get to all of the fasteners.

SPEND WISELY

What really counts is that the standard 9-inch rearend is twice as strong as an 8-inch, and that's very important for a street rodder with a high-torque V-8. The much less common 8 3/4- and 9 3/8-inch ring gear rears are typically more massive as well as expensive. But they do not add any practical value to a street rod.

Then there are the Traction-Lok and Detroit Locker limited-slip differentials. In my opinion, only the street rodder who more than occasionally goes drag racing can justify the expense.

THERE'S MORE

Some 9-inch Ford rearend axle housings have a 9 3/8-inch third member and some of those came with four instead of two pinion gears. And yes, some came with nodular iron gear housings. Whew! But again, they are not cost-efficient for a street cruiser.

WHEN ALL IS SAID AND DONE

There's only one 9-incher that's worth a premium price over the standard late 1960s/early 1970s Mustang/Cougar issue—the 1977–'80 Lincoln Versailles. The pinion and driveshaft are centered in the axle housing like our beloved old Fords and they are factory equipped with disc brakes.

GM REARENDS

I hate to nix most anything that is economical in the boneyard search for street rod components, but I have to make an exception when it comes to the "10-bolt" GM back axle. A stock 350 Chevy small-block has more than enough torque to scatter the 10-bolter when you get on it. And no matter what you say there will come a time when you get on it—it's a genetic defect in street rod types!

You don't have to be a drag racer to rue the day you installed a constitutionally weak rearend. This scattering of a Chevy rear gear was not done on a racetrack, it occurred at a street rod activity!

THE RODDER'S DOZEN

When looking for a Chevy rearend for a V-8 powered rod, the admittedly more expensive 12

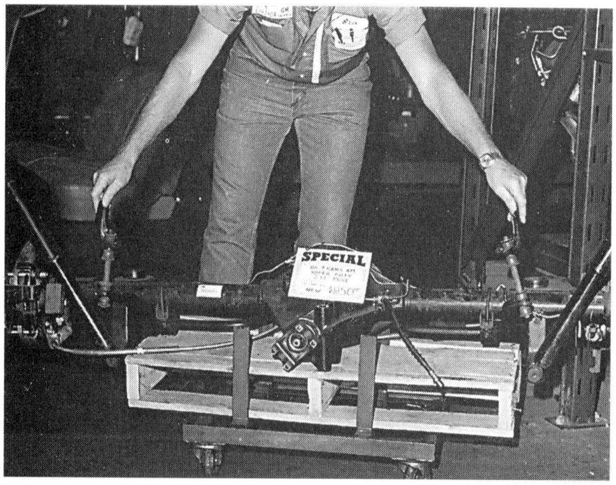

bolt will keep you up and running much longer. Strong Salisbury-design rears can be found under 1970–'80s GM muscle cars. Many are a good fit under early Chevy rod material, but measure your car to be sure.

CHARGE IT!

Just so the MoPar fans won't feel left out, the 1973 Charger rearend is about the same width as many of the OEMs found under 1940s Dodge and Plymouth cars. The original spring pads from the stocker's housing can be recycled, although several neater and more versatile aftermarket installation kits are available.

A QUICKIE

Few undercarriage accouterments are as traditional (and as versatile) as the adaptation of Halibrand Quick Change rearend to a street rod. Some parts and services, however, may not be as readily available as they were in the "old days."

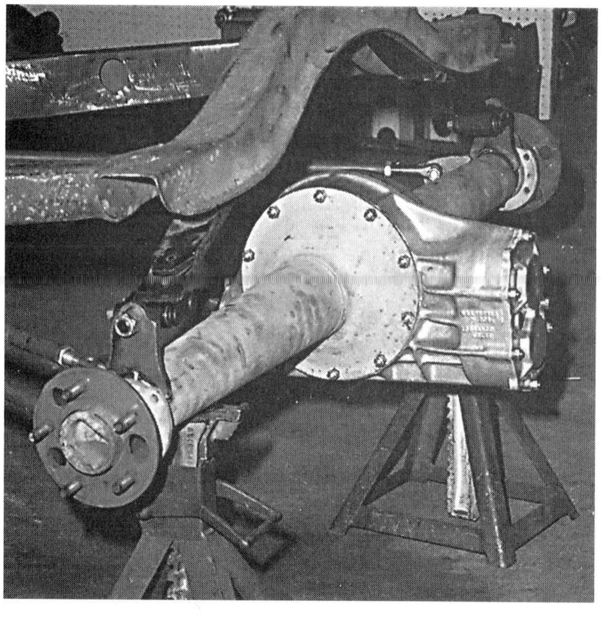

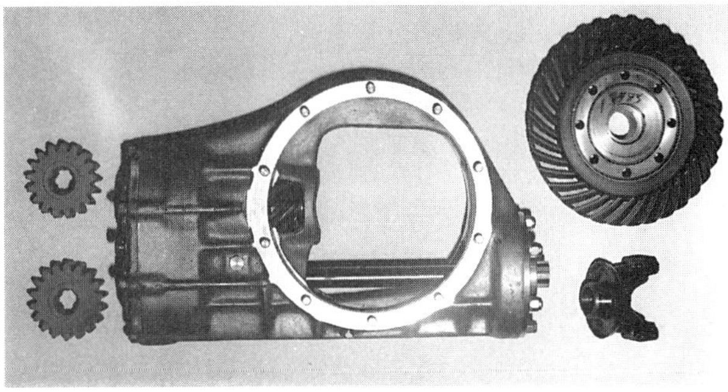

Thank goodness for Gearhead's Quick Change Exchange, 11958 Davisburg Road, Davisburg, MI 48350; (810) 634-5107.

BUYING A USED REAREND

Unless you have accumulated an abundance of free drivetrain parts prior to getting your rod underway, sooner or later the Fat Man's Auto Salvage is going to get a sizeable share of your budget. That's why shopping means more than just looking for the best price. It entails a reasonably careful inspection of the component before money changes hands—and a good understanding of what "guarantees" go with the part.

Obviously an auto salvage manager is not going to permit you to break down a GM 12-bolt differential before you buy it even if you swear you know how to put it back together. He will, however, allow you to pull the inspection cover. With one of the far more popular Ford Hotchkiss types, you should be able to prevail upon him to let you remove the gear assembly for a closer inspection. If you want to do that at home, all you have to do is get him to "brand" the components before you leave the yard. That way, if it has gone south internally, you can make an exchange the next day. At any rate it is up to you to make sure you have good parts before you begin the rearend installation.

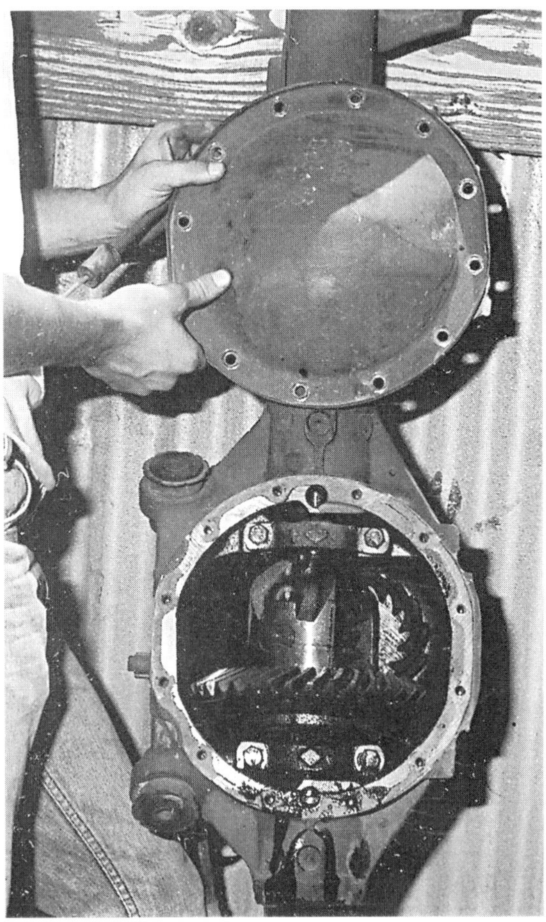

The inspection of a rearend assembly begins with the housing itself. Look for cracks in the cast iron gear case, and for oil leakage at the pinion seal and outside end of the axle shafts. You should also check along the welded seam of the housing for splits or oil seepage.

One of the most important things to look for is a bent housing and/or axle. Half of the cars in an automotive dismantler got there by means of a rearend collision and that could spell bad news for the differential assembly. It'll cost you as much to straighten out a bent axle housing as it did to buy it, so give the prospect a careful inspection. The spring hangers, torsion bars or trailing arms will frequently clue you in if the car suffered a severe impact or was knocked against the curb with great force.

Occasionally you'll run across a rearend that was in a high-performance car. If so, beware of welded-on aftermarket traction bars. All too many have been improperly installed and will have warped the housing.

To check for axle housing warpage, remove the drum or disc brake assembly and rotate the axle flange. Watch for run out. If you are unable to spin the axle by hand there may be some internal galling or binding. Give a listen, as well. If you detect any grinding, pass it up. Finally, take a good look at the brakes and wheel studs on the axle flanges for clues to excessive abuse or damage.

GEAR TALK

The simplest way to get a rough idea of your rod's existing final drive gear ratio is to chock the front wheels, put the transmission in neutral, and jack up the back of the car. Stabilize the rearend on jack stands, mark the driveshaft and hand-rotate a rear wheel one full revolution while you count the number of times the driveshaft revolves. As an example, three driveshaft rotations would indicate a 3.00:1 gear ratio.

Of course, factory gear ratios aren't always nice whole numbers like 3.00:1. They're typically something like 4.11:1. If you get serious about changing final drive ratios, you want your choices to be based on accuracy. The only practical way to know the exact ratio of your gear assembly is to divide the number of teeth on the ring gear by the number of teeth on the pinion gear.

It's a bit messy, but start by draining as much of the lubricant out of the carrier housing as you can, then wipe the ring and pinion teeth dry. Do not use solvent. You only want the teeth dry enough to retain the indexing chalk marks needed to correctly count the gear teeth. Counting can be tricky; do it at least twice.

When you're satisfied your count is accurate, pour fresh lubricant on the dry teeth while you slowly rotate the gears. Or, you can spray the dry teeth with rattle can Cosmoline. The point is, even light surface rust will eventually take its toll on heat treated gears.

RATIO GUESTIMATION

Here's the scenario . . . you've gathered up a half-dozen rear gear assemblies over the years, and you're thinking about changing the gear ratio currently in the car. Most rodders seem to prefer the hum of an engine in the 2200–2600 RPM range when cruising. Beyond that, however, they may not be sure what to expect when switching from Ratio A to Ratio B. Happily, there's a formula that can be manipulated to determine the consequences of a change in gear ratio, rpm or mph, the three most common questions relative to back axle cogs.

Where 336 is a constant:

Tire Diameter x RPM ÷ 336 x MPH = Gear Ratio

or

MPH x Rear Gear Ratio ÷ Tire Diameter x 336 = RPM

or

RPM x Tire Diameter ÷ Gear Ratio x 336 = MPH

(Don't forget overdrive transmissions drop RPM approximately one third.)

There are two other general rules relative to rear gear ratios. The first is: the greater the available engine torque, the higher the gear ratio. (With a 3.50:1 rear end ratio as a midpoint, "high" ratios are numerically lower: 3.25:1, 3.00:1, 2.75:1.)

The second is, the heavier the car, the lower the gear ratio. (With a 3.50:1 rear end ratio as a midpoint, "low" ratios are numerically higher: 3.75:1, 4.00:1, 4.25:1.)

A few words are in order here if you are running a flathead. In the era of the mild, flathead-powered street rod with a manual transmission that actually saw day-to-day transportation duties, the standard 3.78:1 rear gear and the original 6.00x16 tires with their overall diameter of approximately 27 inches was and remains a very practical combination.

If, on the other hand, you've installed an automatic transmission behind that flathead, and you plan to use rear tires that are about 27-inches tall, opt for a rear gear in the neighborhood of 3.90:1.

If you go a little taller on the tire, go shorter (numerically higher) on the gear, i.e. about 4.11:1. Conversely, if you use rear tires that are shorter than 27 inches, go a littler higher (numerically lower), i.e., in the neighborhood of 3.50:1 on the rear gear. The reason, of course, is that the torque output of a mild street flathead with a four-inch stroke is a little on the shy side for an automatic tranny. (A stock 4-inch stroke Merc is only 218 foot-pounds at 1700 rpm.) You gotta make up for it in the final drive ratio.

The above holds true for engines more modern than the flathead, but still in that torque production ballpark—the street rod friendly, small displacement, carbureted V 6s and inline four bangers that preceded the current crop of computerized mini-motors.

DE FACTO DRIVE RATIOS

It's commonly understood that swapping rear tires will, in effect, change the final drive ratio if the new overall tire diameter is measurably different from the old. No mystery, here—larger (or smaller) tires travel a greater (or shorter) distance with each

GEAR INSPECTION

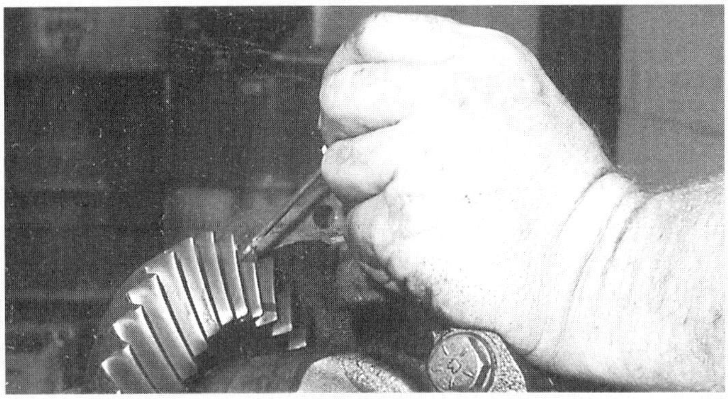

1. Good gear teeth should have sharp, clean-cut edges. Look for rounding off, chips, galling and scoring. Check the "feel" of the oil for bits of metal, dirt, etc.

3. A rearend that was set up properly at the factory should be good for at least 100,000 miles on the bearings. But they do wear out. To check the pinion bearing, put inward pressure on the yoke and roll it over. If it makes noise, or if it feels tight, rough or extremely loose, the bearings have had it.

5. If you are expecting to install a carrier assembly as is, the yoke itself should not have any fore and aft give. If it does, chances are the pinion nut has backed off and lost its preload. When it does that, it will usually tear out a pinion seal. Look for oil leakage here.

2. A rearend that has been neglected in the automotive dismantler may have a little surface rust on the gear teeth. Any rust is too much. The heat treat goes away with rusting and if you try to run such gears the teeth are likely to pit and erode.

4. To check the backlash, hold the pinion solid and rock the ring gear back and forth. There should at least 0.004-inch play. Note the color and consistency of the oil in the assembly. Rearend oil should be changed every 50,000 miles, but it's a rare owner who does. The constantly circulating oil should be as clean as that in the engine.

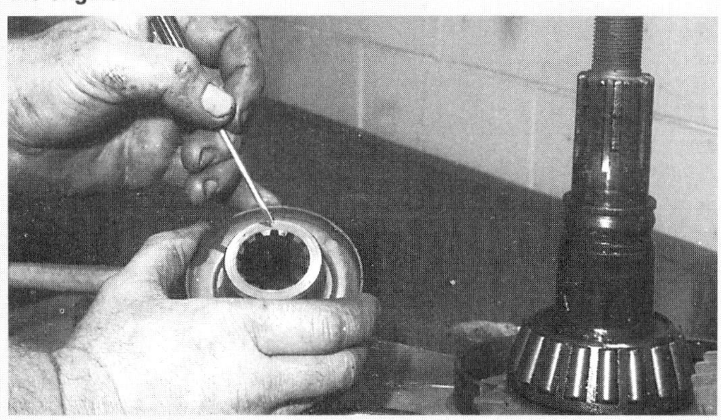

6. Finally, there is sometimes an unpleasant moment of truth. This yoke is worn out. The spline is elongated from slipping back and forth. Replacement of damaged parts is often prohibitively expensive.

revolution. It's because of the tire's rolling radius, and that is empirically determined by measuring the vertical distance from the exact center of the axle to the ground. (The tire must be inflated to the appropriate pressure, but wheel diameter doesn't make any difference)

If you're thinking about switching to larger (or smaller) rear tires, remember you're going to change your final drive ratio. Variance can be calculated with this formula:

Rolling Radius of Old Tire ÷ Rolling Radius of New Tire x The Old Ratio = The New Ratio

My coupe, for instance, has a 3.23:1 final drive ratio and a rear tire with a 12.5-inch rolling radius. If I wanted to swap up to a slightly larger tire with a 13-inch rolling radius

12.5 ÷ 13.0 x 3.23:1 = 3.10:1 (My New Ratio)

That's a significant enough change in tire size to cause my speedometer to read incorrectly. In this case—because it's a larger rear tire—it will read low. (It would read higher if the new tire were smaller.)

CV U-JOINTS

A street rodder had installed a Lincoln Versailles rearend in his street rod, and as he was about to get his driveshaft tailored, he noted with concern that the Versailles originally had a constant velocity (CV) universal joint. Not to worry.

The premium U-joint angle in a passenger car or a similarly configured street rod is one to three degrees. A greater angle will increase U-joint wear and vibration. What the CV U-joint in the center of a driveshaft assembly does is allow the drivetrain designer a little more latitude in cases where the angle is more severe. Conceivably, this could arise in a street rod or pickup that sits very high. However, the "conventional" street rod in which the builder has not gone to any suspension extremes will rarely need a CV joint. A simple balanced driveshaft with yokes that match the transmission on one end and the differential pinion on the other is all that is needed. Nevertheless, all Versailles rears come with the CV flange, and even if you plan to run the relatively tall ratio factory-equipped ring and pinion assembly, it has to be swapped out.

DRAIN-O

Most Ford rearend housings do not have an oil drain plug (above), but one is handy if you like to play with gear ratios. Although the plug can be installed when the rearend is in the car, it is easiest when it is out. Drill a hole in the deepest part of the housing with a "K" bit, and tap for an eighth-inch pipe plug.

MR. CLEAN

But don't stop there. Cut the caked grease behind the bearing supports and down in the carrier area with spray can carb cleaner, then wash out the entire axle housing with hot soapy water. You never know where shrapnel may be lurking!

THEN LUBE IT!

Normal rearend operating temperatures can run as high as 375–400 degrees. Furthermore, the ring and pinion gear teeth are subject to loads exceeding 50,000 psi. As with any lubricant, the purpose of rearend oil is to prevent metal-to-metal contact by providing a thin film between the teeth. Standard 90W with a good film strength additive is just about right for a freshly rebuilt unit. However, if your back axle has seen a lot of service, you may want to consider a slightly heavier grade.

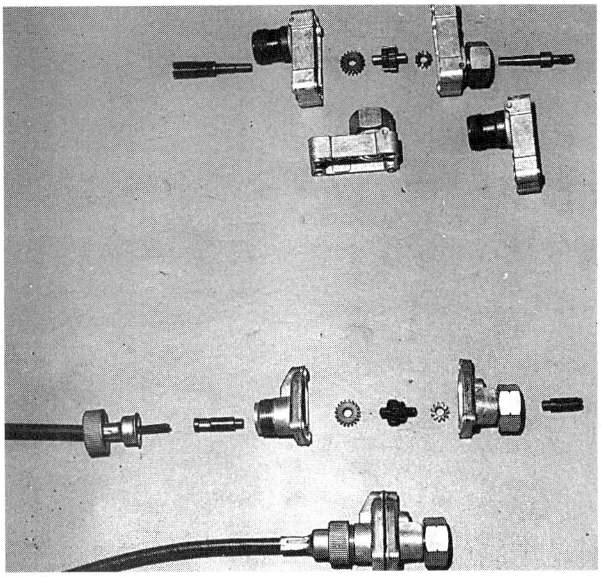

THE FORGOTTON GEAR

When changing rear gears or opting for a taller tire, remember that the combination will affect the speedometer reading from 5 to 15 percent. American speedometers that are cable-driven from the transmission are typically calibrated at the rate of 1000 revolutions per mile.

A quick and cheap fix is to change the color-coded plastic drive gear in the transmission. It's usually a simple job as the retainer is only fastened to the case with one bolt.

If the odometer is too slow, exchange the existing gear for one with one less tooth. They're dealer available and inexpensive. Vice versa if the odometer is too fast. You may have to swap gears more than once to get it right, but if a three tooth change won't do the job, it's time to head out to the speedo shop for a ratio adapter.

VIBE ANALYSIS

Excessive vibrations emanating from your back axle are usually from worn shocks. Check your tires for "cupping" first. Aside from that, the most common vibration problem is wheel/tire imbalance, exacerbated by the street rodder's desire to keep those unattractive lead weights out of sight.

Strange noises in the rear end can be the result of a dry differential assembly. If it's not too late, first put in fresh oil. Then jack both wheels off the ground and spin each as fast as you can to lubricant those thirsty gears.

Impending rear axle bearing failure can also be the cause of noise and/or difficult-to-locate vibration. Don't wait too long before checking it out!

SOME DRIVESHAFT THOUGHTS

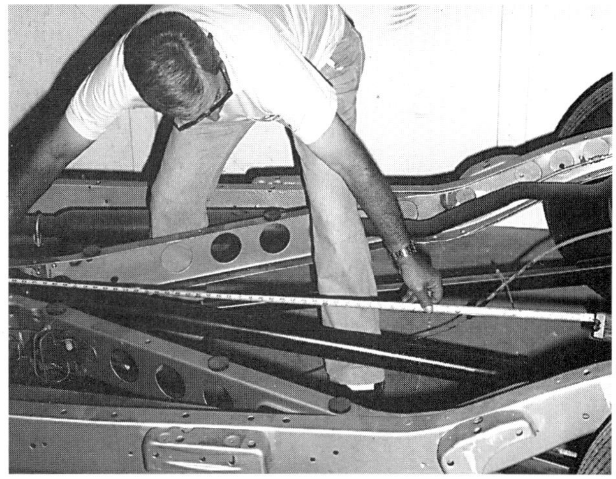

As previously stated, the preferred universal joint angle in a conventional wheel base street rod is one to three degrees. Simply chopping the driveshaft and fitting it with yokes that match the transmission and the newly installed rearend is not enough. Always have a modified driveshaft precision balanced. In fact, even if you are so lucky as to find a stocker that fits perfectly, don't trust to factory specs—get it balanced at a quality driveline service shop.

Neatly storing those extra rearend gear assemblies don't have to be a problem. Simply trace a rearend gasket on a piece of plywood and cut out the hole with a saber saw. Square up the remaining pieces into a box, nail 'em together and drop the gear in.

MAG EVERYTHING!

Finally, please remember that the street rodder's passion for long-in-the-tooth engine and drivetrain components invariably places him in a precarious position. It MUST be assumed that all of them have endured abuse. Never install any old component without first acquiring a Magnaflux check. If you trust to luck, it's bound to be bad!

YOKE SWAPPING

I don't know where you get your rearend gears, but one of Southern California street rodders' favorite places is the local "ecological centers." These may not exist in other parts of the country, but they beat the prices of most conventional automobile dismantlers with no exchange. And, if you are energetic enough, you can build a good collection of gears over time.

Trouble is, FoMoCo 9-inch/28-spline gear carriers come with all sorts of universal joint companion flanges (commonly called "yokes.") You might luck out, and the rear gear you've been waiting for has the right yoke for your driveshaft. But then again, it might not.

I'd hate to tell you how many street rod builders don't realize that changing the yoke is a lot more complicated than merely removing the pinion nut with an impact wrench, and running the new yoke down as tight as it will go. Do it that way and chances are better than even that you will screw up the precision preload setting on the pinion/ring gear assembly. Driving a rod with an improperly installed yoke will give new meaning to the old saying "Humming down the road."

Swapping universal joint flanges is a delicate, time-consuming operation, but one that can be done with tools ordinarily found in the home shop. Follow the bouncing ball as Terry Berzenye of Specialized Street Rods (15542 Computer Lane, Huntington Beach, CA 92649; 714.894.9947) shows us how. Your local Ford dealer will have all the small parts needed for a yoke swap.

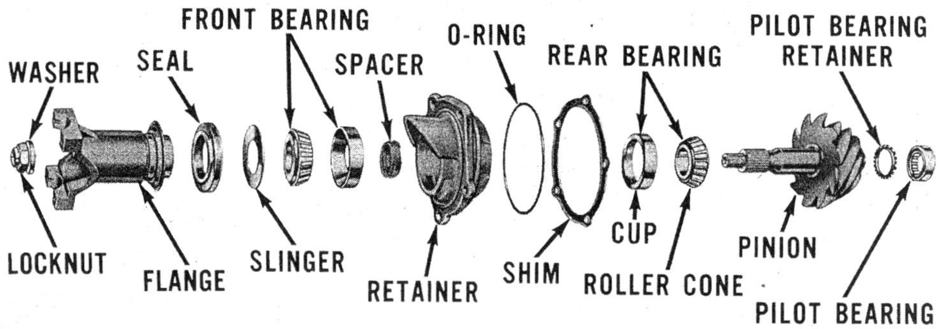

PINION AND BEARING RETAINER

1. Just one of the reasons the FoMoCo 9-inch rearend is so popular is the removable gear carrier. After gaining a bit of experience, a street rodder can change the final drive ratio in a Saturday. But unless he wants to modify the existing driveshaft, the rearend flange has to match the flange that was on the previous gear. The first step is to remove the 3/8-inch bolts that secure the drive pinion retainer to the carrier housing. It is a snug fit, so a few gentle taps with a plastic or brass tipped hammer is in order. Take care not to damage the shim as it can be used again. The thickness of the shim is a critical, factory-determined dimension. Extreme care must be taken not to damage the mounting surfaces of the retainer and carrier.

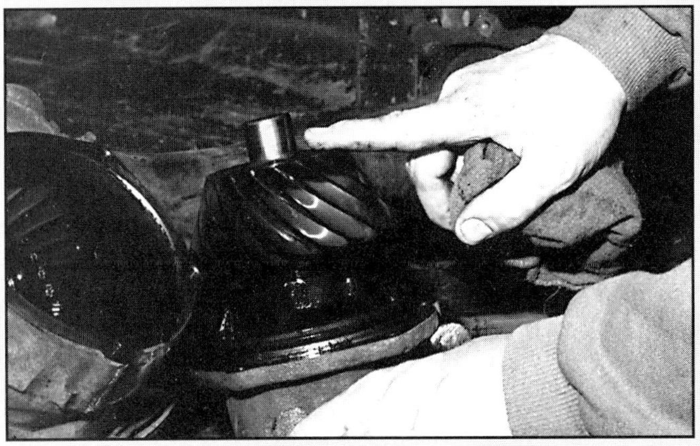

2. Inspect the end of the input shaft bearing surface. If it is pitted or grooved (because of inadequate lubrication) it will need to be replaced. FoMoCo rears are long-lived and tough, though. Chances are good you can continue without fear.

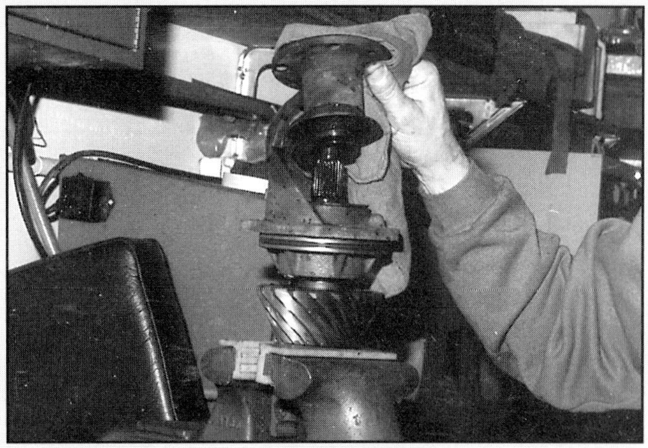

3. Secure the pinion assembly in a vise after taking precautions to protect the end of the input shaft from vise jaw damage. A piece of hose is a handy protective sleeve on the pinion pilot bearing surface. It is all right to take the flange nut off at this point with an impact wrench because you will be replacing the pinion bearing spacer, commonly called the "crush collar."

4. A crush collar can only be used once and must be replaced every time the companion flange is removed for any reason.

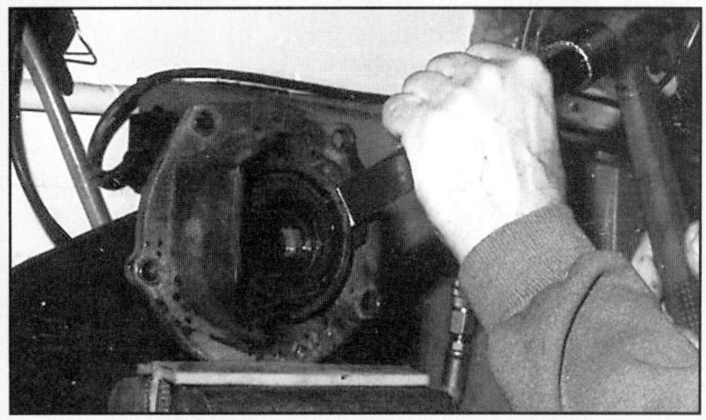

5. There is a judgment call, however, as to whether or not to replace the front and rear pinion bearings and races. Sometimes they last for years. Ah, but sometimes, and for a variety of reasons, they don't! To be on the safe side, replace 'em. Knock out the pinion bearing race first.

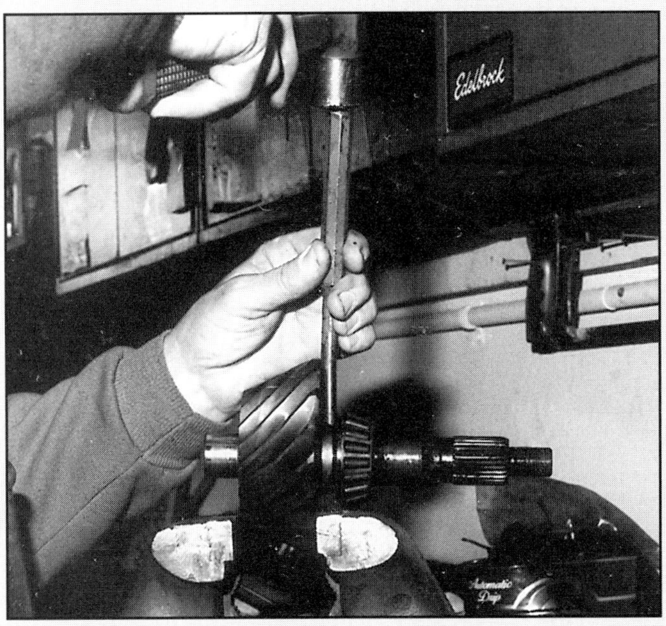

6. Next remove the rear pinion bearing. Hopefully, you've learned the gentle art of persuasion with a hammer; don't damage anything.

7. Remove the original pinion seal and front pinion bearing. If you replace one component, it is standard procedure to remove and replace all accompanying components.

8. Reassembly begins with the replacement of the new front and rear races. After the new races are installed, make sure they are seated in the retainer by trying to insert a 0.0015-inch feeler gauge between the race and the bottom of the bore. You will have to scrounge around for a few pieces of aluminum rod and pipe to make the "precision drive tools." With the new rear pinion bearing in place, slip on the pinion retainer and front bearing.

9. Drive the seal into the pinion retainer. Keep it straight, and tap it in gently until it bottoms out.

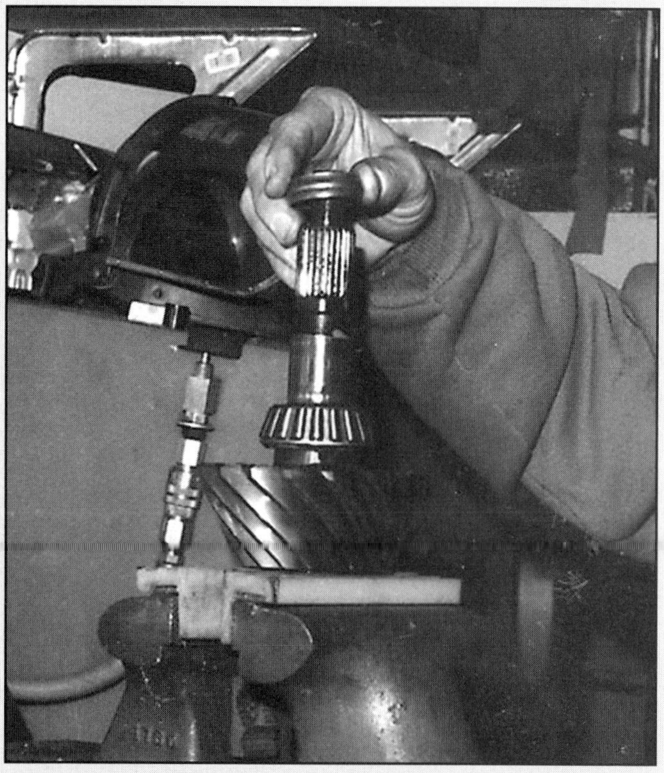

10. Next, slip the new crush collar (pinion bearing spacer) on.

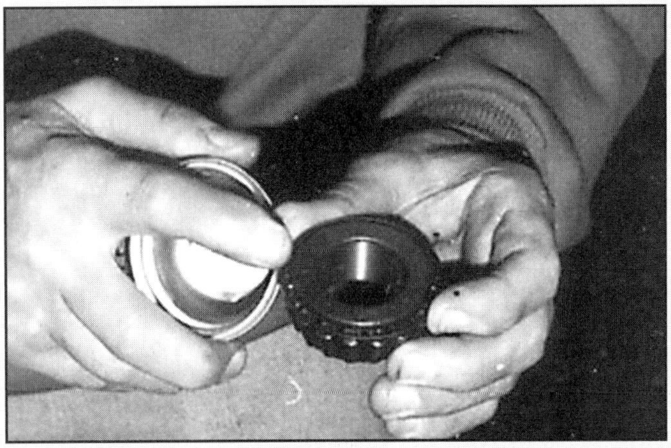

11. Pre-lube the bearings before you put the pinion retainer on the pinion shaft. (The retainer already has the front bearing and washer and seal installed.)

12. Don't put on the new flange before you check it for excessive wear. Look for wear grooves, they be could be a possible leakage point. Pre-lube the flange so it doesn't damage the seal when it slides on.

13. Put some Perma-Tex on the seal that goes into the pinion retainer. Just one more precaution against those pesky driveway leaks.

14. Now you can slip the new yoke (companion flange) on and run down the nut.

15. You can use an impact wrench to draw the flange down snug. Most impacts won't tighten enough to crush the collar, but hold the flange, and tighten only until you feel the play go out of the pinion retainer.

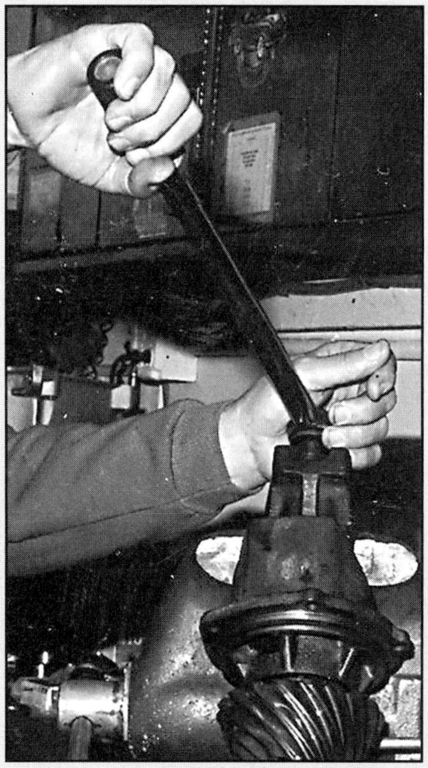

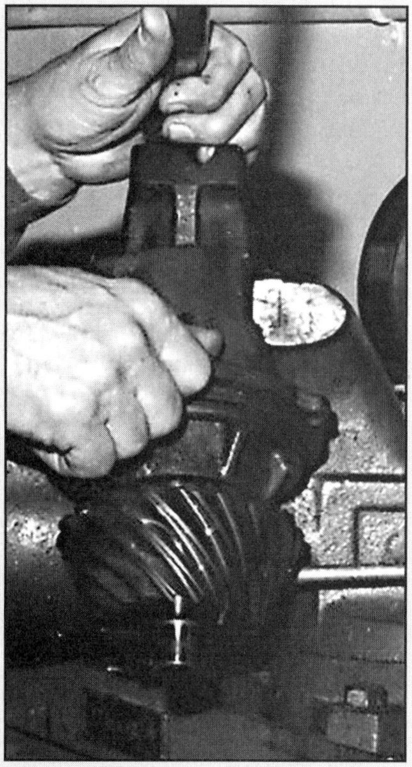

16. As you torque down the nut, move the retainer back and forth until you feel a slight resistance. Do not exceed 175-foot pounds. Tighten the pinion shaft nut a little at a time until the proper preload is established.

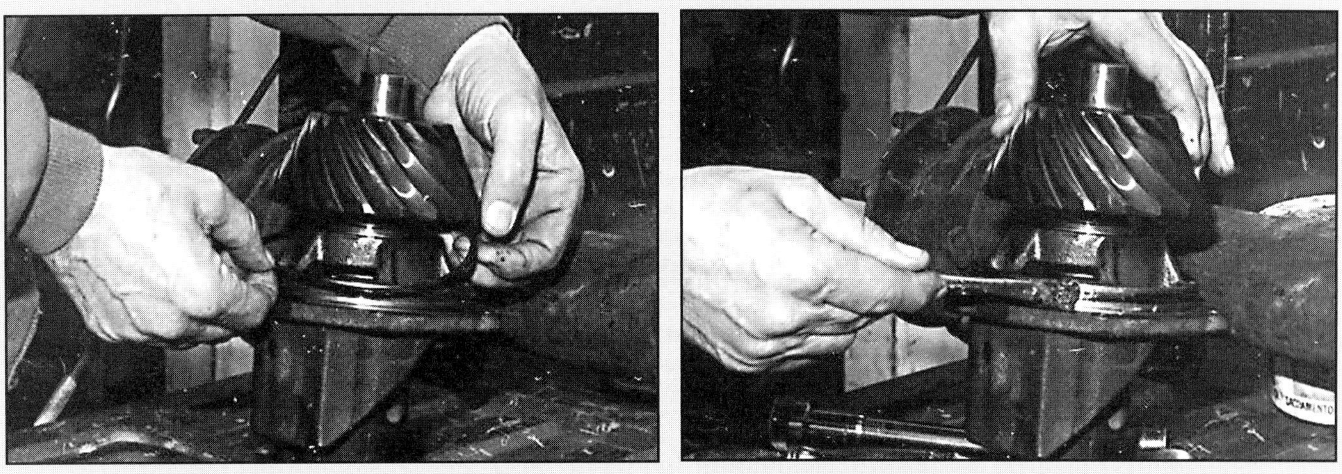

17. Install a new rubber O-ring seal in its groove in the pinion retainer. Be careful not to twist it; put a dab of axle grease on it.

18. No problem reusing the original shim, in fact, it is preferable to use the same one that was originally used.

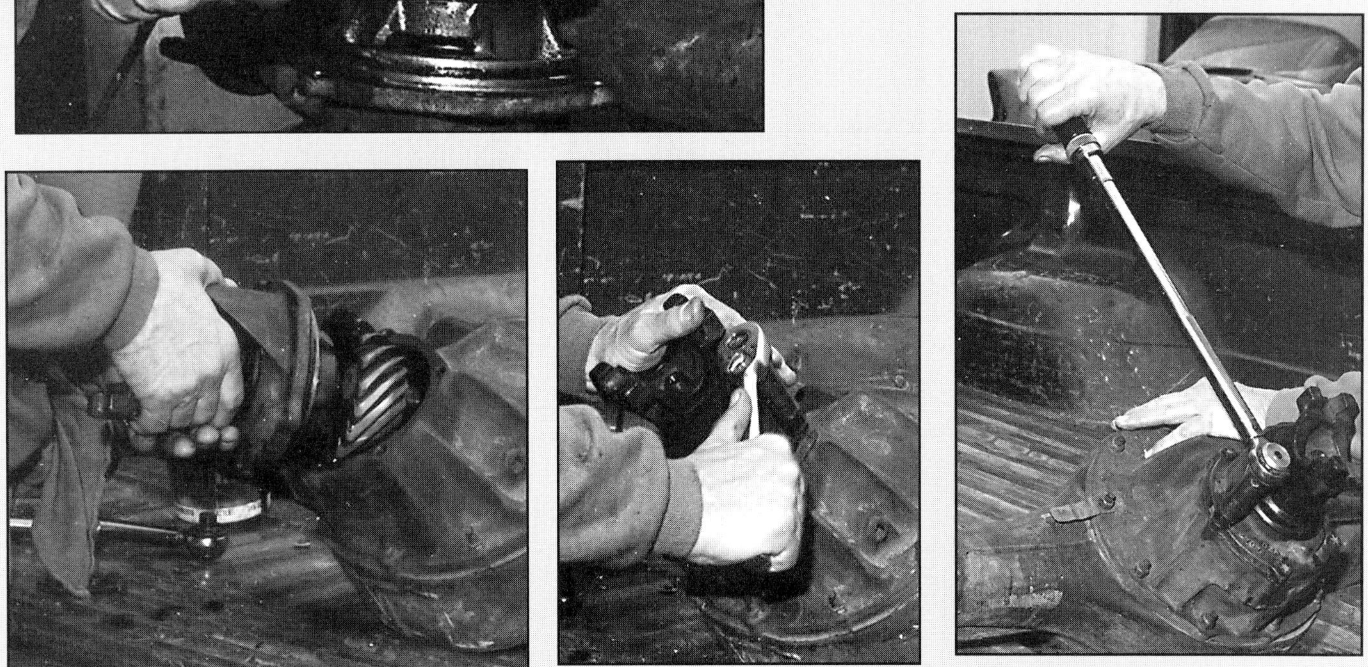

19. Slip the pinion assembly back into the housing, snug it down with a ratchet, and then torque it to 50 ft.-lbs.

Chapter 4

SUSPENSION TIPS & TECHNIQUES

TRADITIONAL SUSPENSION

There's an old tongue-in-cheek quote that goes like this "Progress is fine, but it's gone on for too long." In the eye of many street rod traditionalists, some rods just don't look "right" if there is too much technological progress up front and in your face. Ford-based street rods from Fad Ts through the 1934 models simply demand a basic "semi-traditional" buggy-spring front suspension, with or without fenders—all the hardware is exposed to public view.

Simply improving some original suspensions is often a good way to go to keep a veteran street rod up and running until major modifications can be undertaken. One such case is the "knee action" suspension in a late 1940s resto-rodded Chevrolet. Often, just rebuilding the front suspension restores adequate ride characteristics, but adding modern tubular shock absorbers provides a significant improvement in handling. The latter is simply achieved by adding brackets to the frame and lower A-arms.

Kanter Auto Products of New Jersey (www.kanter.com) is one of several companies that sell most of the stock front suspension components and all of the street rod chassis shops have shock mounting kits in one version or another. There is one caveat—always use new springs (leaf or coil) when reworking an old car's suspension.

TRUE TO FORM

If you plan to hang a new suspension crossmember on an original chassis, remember a lot can happen to a sixty-plus year old frame even if it looks perfect to the untrained eye. The place to begin, of course, is with a spotlessly clean, sandblasted frame and a big magnifying glass. Check for stress cracks in all areas where the factory attached one original frame member to another and where the steering gear and suspension components were attached. Small tears can be welded at home, but loosened rivets can mean a "tweaked" chassis.

A chassis in need of serious repairs is best rehabilitated on a body shop frame alignment machine where it be checked for "trueness" before work begins as well as after surgical

intervention. Any well-equipped body shop can handle the task. It's not cheap, but a true frame is essential; not only is it the anchor for all suspension components; it is the foundation upon which the body must be placed!

HELP FOR THE NEEDY

It never hurts to add stiffening to an original chassis slated for street rod service. There are universal crossmembers available, and with one as a start, the builder can often design and fabricate the rest of what he needs for what he has.

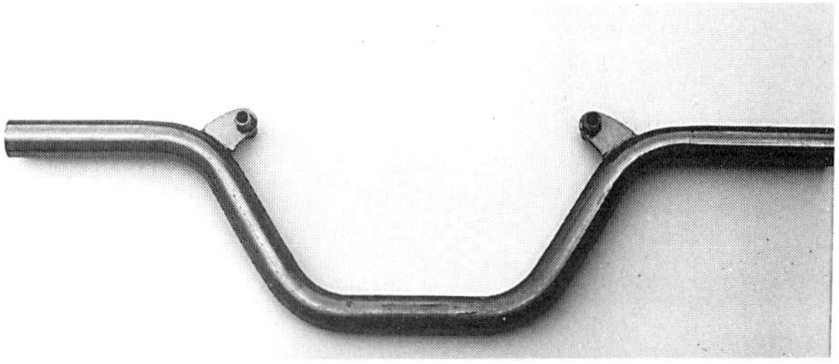

Boxing plates can be particularly useful in and around the front suspension, especially if a high torque motor is in the future. Flame-cut plates are available from several chassis shops.

UPGRADING THE RARE BIRD

Affordable street rodding material in the mainstream of popularity has all but disappeared. Consequently, would-be rod builders have had to expand the boundaries of "popularity." Engine, transmission and rearend swaps are normally straightforward, but updating the steering and front suspension of a car from unexplored territory and for which kits have yet to be designed can be a daunting task for the novice. Not to worry, you simply have to provide your professional resources with enough information for them to make practical suggestions.

The first step after removing the front sheet metal and engine is to tow or trailer the remainder to the quarter car wash. Using their high-pressure engine cleaning facilities, blast the forward end of the chassis and existing front suspension. If hauling your car around is not practical, do the best you can at home with Gunk and hot, soapy water.

Next, paint everything from the lower firewall forward with a few rattle cans of "hot rod gray" primer. The primer doesn't hurt anything, and it cleans off easily with lacquer thinner. It makes grungy old automotive running gear much more photogenic.

Then photograph the frame, stock steering, suspension, and steering column from all angles.

Measure the axle, front track, wheelbase and whatever stock drag links, tie rods, Pitman and idler arms that may exist in these uncharted waters. File your information in a folder, and put it in your desk.

Next, send a short letter of inquiry with an addressed and stamped envelope to all the appropriate kit manufacturers who advertise in the rod mags. You'll be surprised at what resources are out there.

MUSTANG II IFS

M-II installation kits are made for most popular street rod material, but not every car built in the thirties and forties. Don't let that stop you non-Ford guys, however. Contact manufacturers for suggestions as to which of their existing products could be adapted to your car. Pictured is a kit offered by Heidt's Hot Rod Shop.

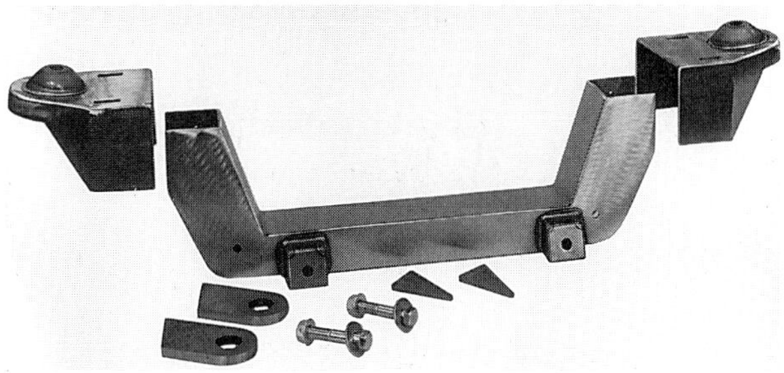

"Side sway" varies from car to car depending on ride height, the type of rear suspension and the stiffness of that suspension. Even at moderate speeds, the side load on the tires is transmitted from the axle to the frame through bending or twisting the springs. Often some kind of locating device to prevent axle deflection (which is what causes tires to rub wheel wells) is needed. In other words, the axle must be laterally connected to the frame. The Panhard rod is the simplest locating device.

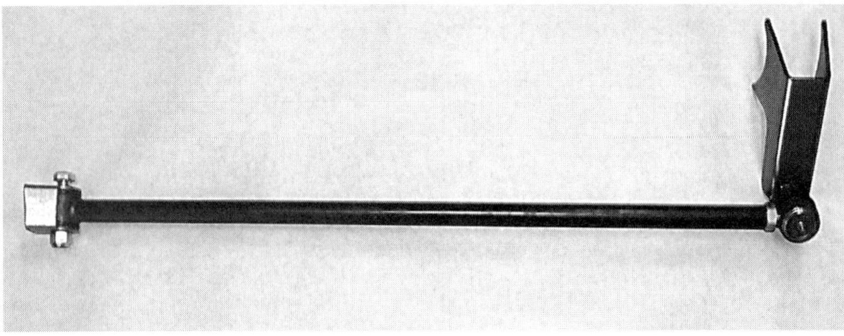

A Panhard rod is more rigid than the springs, and consequently able to handle most of the lateral load. The most basic was the one used on the 1947–'48 Ford. It was a belated, but major improvement on the transverse ("buggy") spring suspension. The modern SAC Hot Rod Products' version is pictured. The Panhard rod must be free to move about and should never interfere with rearend suspension movement. It runs across the width of the car with one end pivoted on the axle and the other end pivoted on the frame. At ride height the Panhard rod must be horizontal; otherwise it will induce too much lateral axle movement, a potential cause of structural failure.

WATTS LINKAGE

Watts links in various forms are also used to locate live axle suspensions. The preferable design is to mount the vertical pivot in line with the center of the axle housing and two horizontal bars from there to brackets on each side of the frame.

NEW AGE ANTI-ROLL BARS

Often called "anti-sway" bars, these devices resist body roll and the resultant weight transfer by transferring load from an inside wheel to the outside wheel during hard turns. The kit pictured below and installation is marketed by SAC. Hot Rod Products of Anaheim, California.

SPLITTING THE WISHBONE

Although the original early Ford transverse spring radius rod assembly is perfectly adequate in a stocker, even when it's not needed to provide clearance for a street rod power train swap, "splitting the wishbone" allows individual adjustment of each end of the axle fine-tuning front end alignment. (Kits are available from most chassis shops to minimize the angular spread.)

HONE THE BOSS

When planning to install a new set of kingpins in an early Ford front suspension, the spindle bosses should be align-honed (at the machine shop) for a precise kingpin fit.

HYDROGEN EMBRITTLEMENT

This is a long-winded topic, but here's a short, to-the-point tip: Critical suspension and steering components should not be chromed because of the adverse structural/chemical reaction induced by the electroplating process.

If you have acquired such components along the swap meet trail before becoming aware of the potential danger, they may yet be salvageable. First, however, they must be stripped of the decorative chrome. Unfortunately, few chrome platers are willing to risk contaminating their tanks during the de-chroming process, so keep that in mind when looking over used parts.

But that's not all, the parts have to be oven-baked to remove the free hydrogen molecules. Powder painters have the facility to do so. (And that's what should have been done in the first place!)

TEEING OFF

I've talked about this before, but here are a few tips worth repeating. A retired Ford Motor Company suspension engineer brought the concept to my attention . . .

When the "T-bolts" that secure the upper control arm to the crossmember of the stock Mustang II IFS (or its aftermarket clones) are vertical, they are placed in shear. As such, they are more likely to slide from load. A horizontal position was strongly suggested. This led me to have gusseted uprights welded to the upper section of the crossmember in both my street rods. The 5/16-inch upright plate is 6.5-inches long and 2.75-inches high with gussets as needed in the readily identifiable stress points.

In my case, the crossmembers had been installed as per the instructions that came with each kit. Bud Matthews (the fabricator) marked the top hat at the centerline of the cross-shaft and also at the centerline of the T-bolts. He then undid the T-bolts and rotated the cross-shaft 90-degrees. With the T-bolts and an eighth-inch shim now parallel to the top hat, the position of the cross-shaft was established as per the marks on the top hat and original T-bolt holes. After verification, the pre-cut bracket and gussets were welded in and the bolts tightened to 95 foot/pounds torque for a secure bite.

BETTER SAFE THAN . . .

I've already mentioned recycled "senior" suspension and driveline components should always be Magna-fluxed. I want to extend that admonition to "mature" load-bearing components such as springs, spindles, axles, driveshafts, and so on. Even after as "few" as twenty years of wear, an indeterminate amount of abusive use can take the starch out of the best steel.

Generally, a minimal reduction in the free height of independent front suspension coil springs will lower the car's nose without preventing an appropriate caster and camber realignment. Problems always develop when you go beyond

moderation. Extreme lowering usually requires extreme means to regain proper ride and handling characteristics. You will be surprised to see what the removal of only one loop accomplishes.

CUT WITH CARE

Under no circumstances should a cutting torch be the weapon of choice when modifying springs. Heat changes the temper of the metal and leads to premature failure. You can safely remove a single loop of a front suspension coil spring to lower ride height, but use a pneumatic cut-off tool and deburr the sharp edges. Always wear eye protection.

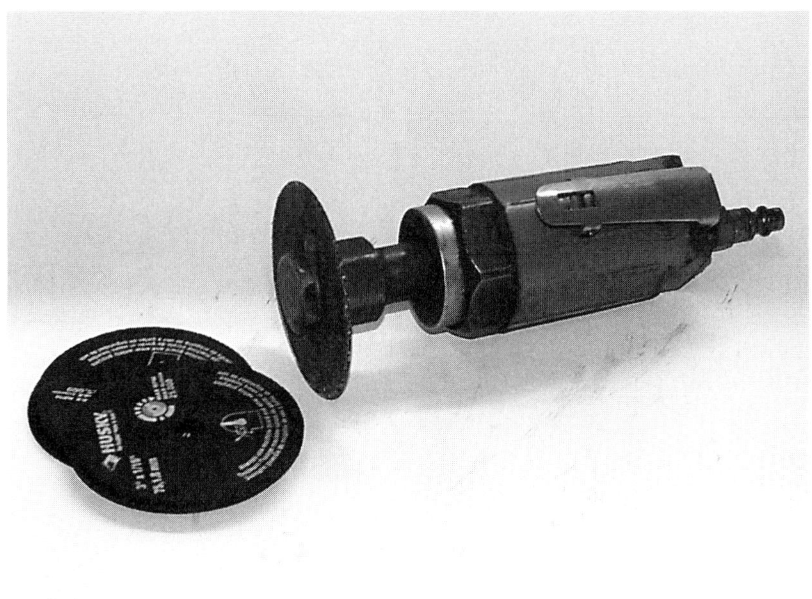

TO CUT OR NOT TO . . .

Independent front suspension coil springs are not commonly wound from tapered wire, but if yours are, don't cut them. Replace them. The truth is, replacement is always the preferred modification. Auto parts houses usually have an extensive catalog library, and softer or stiffer substitutes are often available.

Another option is a custom-wound spring. Most metropolitan areas have a spring rebuilding shop that caters to the trucking industry. It never hurts to discuss your particular suspension needs with them. Professional modifications made at their convenience can often save you money.

'VETTE SPRINGS

One of the class street rod independent rear suspension swaps is the installation of a Corvette IRS with springs a la 1978 models. However, if there is a ride height problem, try the lightweight fiberglass replacement springs manufactured by TRW and sold by Eckler's Corvette Supply, 5140 S. Washington Ave., Titusville, FL 32789. (e-mail: sales@Ecklers.net)

CRUSIN' COMFORT

There's a lot of subjectivity in street rodding, and you won't find more than in one's assessment of "ride." It is largely a matter of comfort, and the standard is whatever your backside is accustomed to in your everyday means of transportation. If you are accustomed to a new Lincoln Town Car, your expectations are going to be high.

ANGLE OF THE DANGLE

With regard to shock absorber angle, 30-degrees

from a vertical plane through the upper mounting point is the happiest medium. Generally speaking, less than 30-degrees will result in a stiffer ride, whereas an angle greater than this will give a softer ride. Also, the closer you mount the shock to the wheel, the more efficient it will be. This location provides maximum shock travel and stiffness, regardless of what the stiffness (damping) rate may be.

HOW SHOCKING!

Always check shock-to-wheel and shock-to-chassis clearance over the full range of suspension travel. Beyond that, when the chassis sits at its normal ride height, the shock should be at its midpoint of travel. And make certain the shock never reaches its extension limit before the wheel reaches its maximum travel, in both compression and rebound. Instant shock damage is a certain result.

U-JOINT ANGLE

The plane of the front and rear universal joints must be exactly parallel, and the angle of the trans output shaft and driveshaft must equal the angle of the driveshaft and rearend pinion yoke. The angle should be as small as possible—no more than 4 degrees at rest and 20 degrees when the suspension bottoms.

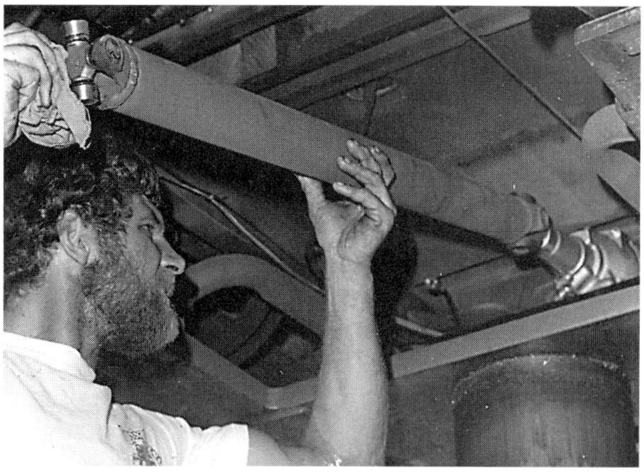

Although a universal joint with a Zerk grease fitting is not as strong as a solid universal joint, the solid type offers no particular advantage for a street rodder who is only a spectator at a drag race. Ease of maintenance is more important.

DRIVESHAFT LENGTH

Determining the proper length of the driveshaft seems simple enough, but costly mistakes have been made. First, slide the yoke onto the transmission output shaft until it bottoms out. Next measure the distance from the centerline of that yoke's U-joint

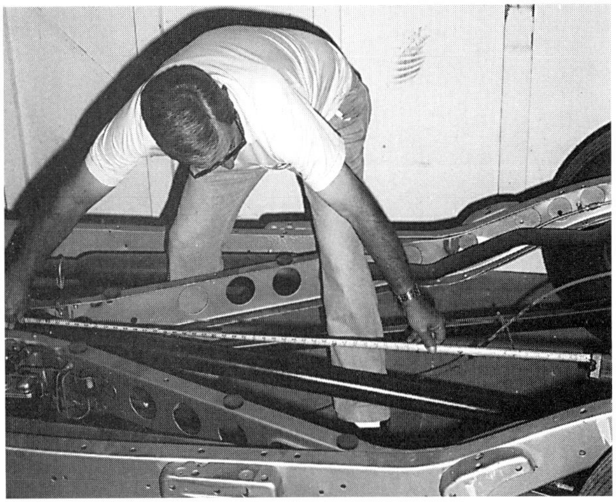

to the centerline of the U-joint on the rearend companion flange, then subtract two inches. When the new driveshaft has been made, it must be straight within a tolerance of 0.0005-inches, and it must be balanced.

SWIVEL HIPS

Avoid spring bind and ride height problems when installing a transverse spring with Posies adjustable

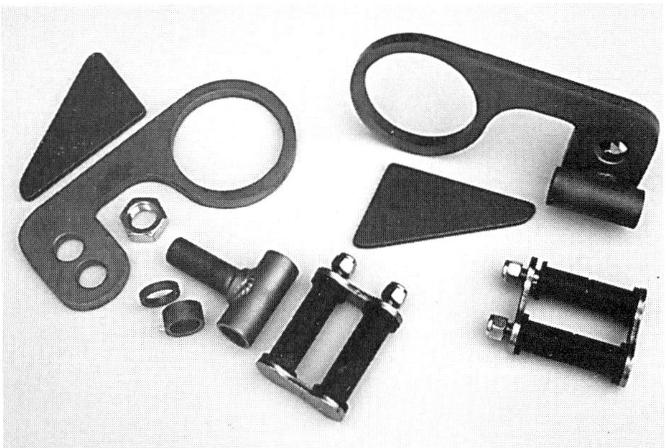

spring perches. The perches swivel in the hangers allowing the pinion angle to be accurately set when the finished car is on all fours.

WHEELS AND TIRES

When you get that scheduled front end alignment, tell the technician in advance that you are after the absolute maximum in tire mileage and fuel economy; he will dial in a tad more positive caster for stability.

Don't approach your tire inflation with a "by guess and by golly" attitude. Pump them up to the top of the range indicated on the tire, and evaluate the wear pattern. If the inner rows of the tread seem to be wearing faster than the outboards, back off by a pound or two. However, if the wear patterns look

good, increase the pressure a pound or two. A constant surveillance of tire wear patterns should become an ingrained habit.

After truing, all four tires should be rebalanced, and it is important that they be balanced dynamically. Unfortunately (from a cosmetic point of view), balance weights are most effective on the outside of the wheels. (That's just one reason I like my painted steelies!)

New wheels and tires are always a street rodder's delight, but there are many other considerations that affect fit and performance. Obviously, the type and style of front suspension plays a significant role. Independent front suspensions (installed with commercial kits or one-off owner-fabrications) present different situations than I-beam axles with or without a drop.

Much depends on the turning radius of the steering gear; some screw in closer than others. In other words, "elbow room" determines the appropriate size wheel and tire for the fendered street rod front end. That's also the case for all the other components of the suspension (and their location)—the shock absorbers, and the type and configuration of the brakes. (If discs are used, how big are the calipers?) You also have to factor in the location of the anti-sway bar, and modified radius rods such as the four link assemblies.

For all these reasons, it behooves the street rodder to establish a baseline. Since you really don't know for certain how your suspension is going to work, there's no sense in dropping a bundle on fancy alloy wheels and the finest tires just yet. Start with plain

Jane tires, stock steelies with a coat of paint. You can upgrade at your financial leisure. The point is to get the car on the road so you can properly evaluate suspension operation and those subjective intangibles such as ride and handling.

Finally, there isn't a hard and fast formula relative to the variance between the overall diameters of front versus rear tires. The Rule of Moderation always applies.

STEP TO THE REAR

As if all the potential mechanical interference on the front half of the car isn't enough, you've still got to contend with the back half. Of course, the primary consideration is the width of the rear axle assembly. Again, don't wait until you are at the Goodstone dealer to find out the newly purchased tire and wheel assembly doesn't fit! Borrow a pair of

wheels and tires that are dimensionally similar to what you think you want. You'll only need them for as long as it takes to evaluate the static clearance and ride height.

When you're running a full-fendered car, it is well to avoid a wheel (front or rear) with a lot of positive offset (the distance from the wheel-mounting surface to the center of the rim). One inch of static clearance is the minimum offset. And, please don't determine the proper size of new tires and wheels for full-fendered cars by a simple visual estimation. Every direct measurement should be noted and discussed with the tire dealer.

It is common knowledge that when tires are mounted, the assembly has to be balanced. After that's done, just to be on the safe side, jack each

wheel off the ground (one at a time), spin it and observe the rotation from the front and rear. If there is a noticeable run-out, it's time to talk to the wheel dealer because vibration or "wheel wobble" can develop.

Front tires that are "cupping" are not uniformly making surface contact with the road. If the tires are properly inflated, the most likely cause is worn-out shock absorbers. If the existing shocks are not worn out, heavy-duty replacements may be in order.

Tires that are cupped may only be good enough for garage "roll-arounds." If in doubt, check with a full-service tire shop to see if they can be "trued" (buffed or shaved to make them perfectly round again).

VIBRATION ANALYSIS

When tracing down suspension vibrations, remember that they travel from the source (road contact) through the chassis or body. Front suspension vibrations are typically transmitted to the steering wheel.

Improper driveshaft angle, U-joint phasing, or worn driveline parts develop their own brand of vibrations, and valid analysis is beyond the skills of the street rodder. In short, checking these components is a job for a driveline service company familiar with the use of a stethoscope and a Reed tachometer. The latter measures the frequency, amplitude, and audible noise of the vibration, and enables the technician to track down the source.

TECHNIQUE: QUICKIE (AND TEMPORARY) FRONT ALIGNMENT

This street rodder had his Fad T-bucket ready to roll, the engine had been fired and the drivetrain was functioning. Yet he hadn't backed out of the garage, because the front end had only been adjusted "by eye" and the closest alignment shop was a number of miles away. Yes, few of us build our cars next door to an alignment shop.

Now, Jerry Kiefer's California Custom Roadsters (www.californiacustomroadsters.com) is the place to go when it comes to T-buckets. Jerry often gets front-end alignment questions. "Many of our customers take side roads to get here because they're afraid to go on the freeway. They always want to know how they can temporary set up the front end. Ninety-nine percent of T-Buckets have I-beam axles and the settings are not even close! Basically, I tell them that a tape measure is their most useful tool in making sure the front end is close to dimensionally correct. We do a preliminary toe-in setting here with a tape and they can do the same until they get the car to the alignment shop."

Jerry Kiefer, (behind the windshield post) and assistant show how to set the toe-in at home. ("Toe" is the angle measured between the tires, looking down on them from above. Admittedly, it's easier to see on the fully exposed Fad T front suspension.) Zero toe is the setting where the wheels are exactly parallel. Toe-in brings the leading edge of the tires closer together than the rear. Toe-out is when the leading edge of the tires is further apart. This basic setting dramatically affects handling characteristics.

The geometry of the steering linkage changes the toe setting as the steering wheel is turned. So you must first center the steering. The distance between the outboard tie rod pivot and a suspension pivot on the frame should be equal on both sides of the car with the steering straight ahead. The toe itself is adjusted by changing the installed length of the tie rod between the pivots. (The clamps are loosened and the tie rod is rotated in the threaded fitting.) Direct measurement between the centerline of the tires is the simple way, but remember tire treads give; they are not precisely accurate. A better way is to jack up the front end and carefully scribe a chalk line on the tire around its circumference. (Make sure the wheel bearings are properly installed with no slop.) That done, turn the tie rod in the desired direction to obtain the necessary dimensions. The rule of thumb for toe-in on the early Ford front suspension is from a sixteenth to an eighth inch. This is only a temporary adjustment, however. A full precision front-end alignment (camber and caster as well as toe-in) is in order before you do any serious driving.

TECHNIQUES: WHEEL BEARING PACKING

It's one of those maintenance tasks you essentially know how to do, but you've avoided it. Not because it's difficult, but it's kind of "yucky" as the bubble-gummers say. You don't just get your hands dirty, you get them sticky, and you have to work to get them dinner table fresh.

As a consequence, you may rush through the task a bit more quickly than you should. And that's not good. Packing wheel bearings is one of those jobs that are basically simple, but critical. If you're lucky when the results of a careless wheel pack gives out, you only hear a grinding noise and erratic steering. If you're not lucky . . .

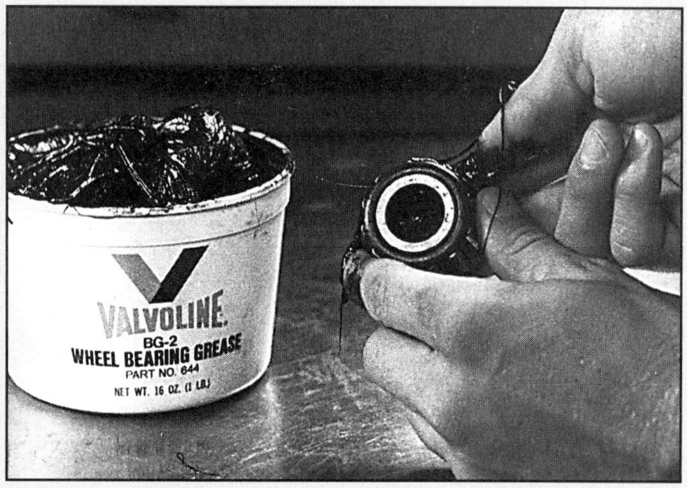

After you've safely moored the car and dropped the wheel, remove the grease retainer and the inner bearing cone and roller assembly from the hub. Clean the inner and outer bearing cups with solvent and look for scratches or pits. Assuming they're not worn or damaged, soak the new grease retainer in light engine oil for 30 minutes before installation. Thoroughly clean the inner and outer bearing cones and rollers with solvent and dry them thoroughly. Do not spin the bearing dry with compressed air. Inspect the cone and roller assemblies for wear. The cone and roller assemblies and the bearing cups should be replaced as a unit if damage to either is apparent. Clean the spindle and the inside of the hub with solvent to remove all old lubricant. Then cover the spindle with a clean cloth, and brush all loose dust and dirt from the brake assembly. Pack the inside of the hub with wheel bearing grease only until the grease is flush with inside diameter of the bearing cups.

Professional mechanics have a special tool for repacking, but few street rodders have such luxuries. They have their hands, however. When repacking by hand, work as much lubricant between the rollers and cages as possible. Then coat the cone surfaces with a light film of grease. Place the inner bearing cone and roller assembly in the inner cups and install the new grease retainer. It is unlikely you have the factory recommended tool, but generations of street rodders have installed retainers with the gentle persuasion of a lightweight brass hammer. Just be sure that the retainer is evenly seated all the way around.

Install the hub and rotor assembly on the spindle, taking care to keep it centered so as to prevent damage to the retainer or spindle threads. Fit the outer bearing cone and roller assembly and the flat washer on the spindle, and then install the nut. Rotate the hub assembly and torque the nut to 17–25 ft-lbs. to seal the bearings. Then back the nut off one-half turn and retighten to 10-15 inch/pounds (finger tight). If the notches in the castellated nut fail to line up with the cotter pin hole in the stub axle to your satisfaction, see if a thin valve spring shim or two behind the nut gives you a straight shot. Install a fresh cotter pin, and bend the ends around the castellated flange. Tapping on the grease cap finishes the job. Obviously, the weight of the car is a significant factor in suspension travel, but all the formulae in the world won't tell you as much as a 30-mph drive over a pair of train tracks. Some things just have to be empirically determined.

Chapter 5

STEERING TIPS

A WEIGHTY DECISION

Rule of thumb—if the street rod weighs less that 2,600 pounds and the engine weighs less than 600 pounds, chances are better than even, power steering is not needed.

VEGA STEERING

The 1971–'77 Vega Saginaw manual steering box is a long-time favorite and is a good as anything you could want. However, the builder must avoid the temptation to recycle the stock tie rods. Check with street rod component manufacturers for current availability of heavy-wall tubular rods with matching tie rod ends.

Some GM Saginaw power steering units are perceived to be too sensitive at highway speeds. Since GM used Model 808 with only two turns lock-to-lock in mid- and lightweight cars, it's not always the street rodder's imagination. (The fewer turns, the faster the box steers.) A friendly automotive dismantler operator can usually come up

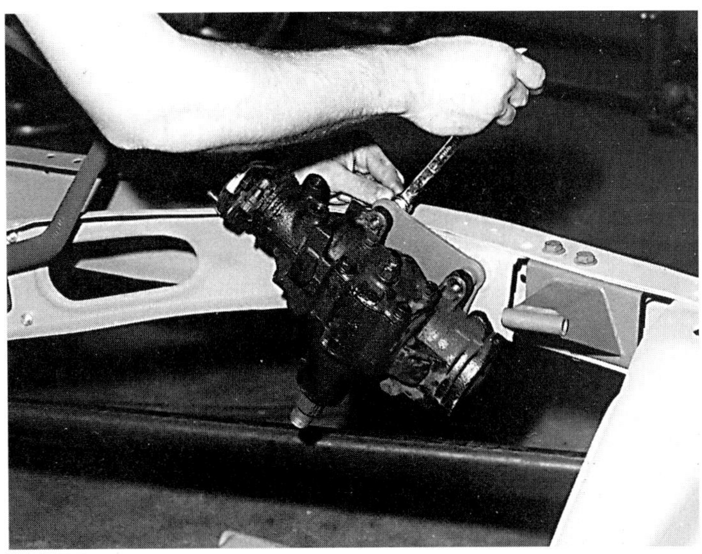

with a full-size car box (Model 800) as slow as four turns lock-to-lock that will interchange.

Model 800/808 Pitman arm sizes and hole diameters are all the same, but there are two different input spline sizes. Simply switching to the appropriate universal joint (stock fabric type or high quality aftermarket offerings) will take care of the mismatch.

Of course, before going to all the trouble of a steering swap, you may be able to get a steering feel more to your liking by simply having the front end alignment rechecked, or lowering the tire pressure a bit.

The Mustang II IFS clone is clearly the most common street rod IFS. Terry Berzenye of Street & Performance (Huntington Beach, CA) convinced me that the petite plastic GM/Delco power steering units used in mid-1990s GM cars with front wheel drive would be just right for street rods in the 2,800 pound range. The range is internally regulated (roughly 200 to 700 psi maximum) and the high end only comes into play during a hard turn. Often, the only significant difference from model to model is the size of the fluid canister.

However, some drivers think it steers a little too fast. Mullins Steering Gears (2876 Sweetwater Ave., Ste. 2, Lake Havasu City, AZ 86406) sells a simple little power steering pump flow valve kit that even a rodder with no experience in automotive hydraulics should not fear.

If you are using a GM power steering pump designed for a Saginaw ball & nut steering in a car with an original or clone Mustang power rack-and-pinion, you are likely to find that the line pressure delivered is a little too much. Study the pump, note where the pressure line comes out of the pump body. A nut is there, and behind it is a spring-loaded bypass or pressure regulator valve. Cut one coil off the spring to get a better match between line pressure from the Chevy pump and the Ford rack-and-pinion. The high-pressure line usually goes to the smaller of the two fitting in the rack-and-pinion housing, but obviously there is a potential fitting problem.

JUST AROUND THE CORNER

Your best resource for Chevy/Ford adaptive plumbing is the local NAPA store. Bring in your pump and rack-and-pinion, and when the counterman has a few minutes to go through his illustrated catalogs, he can usually come up with the correct line length and compatible fitting.

A MOPAR INSIGHT

Here's a steering tip for Mopar rodders—Ken Killen of New Mexico was very pleased with his installation of a Total Cost Involved narrowed, rear-steer Mopar rack & pinion (PN 2932) in his 1936 Plymouth. He kept the original leaf springs and front axle, but replaced the original knee actions with Monroe gas shocks. Interested readers should contact Total Cost directly for details. Have your original front rack measurements at hand when you call.

Rodders wishing to maintain the traditional side steering (with the Pitman arm pointing down) have two acceptable choices. Use the Saginaw 525 box in a car with a gross weight exceeding 2,500 pounds or a modified Vega in street rods styled after the traditional Model T bodied track roadster with a

gross weight less than 2500 pounds. Mullins Steering Gears, (2876 Sweetwater Avenue Suite 2, Lake Havasu City, AZ 86496. 520-505-3032) can take care of the Vega modification.

Whenever the tire size, wheel width, and wheel offset are varied from stock sizes, the factory alignment settings go out the window because the amount of energy and leverage imparted into the front suspension and steering is different. Find a street rod friendly front-end alignment shop with an alert technician who will compensate for the changes rather than blindly following the numbers in an alignment manual.

He who lowers the front end of a street rod has to remember that too much rake reduces caster. (The purpose of caster is to make the steering return to center and keep the vehicle tracking straight ahead without wandering or darting all over the road.) Lowering the front and raising the rear, in effect, tilts the kingpin angle forward. The caster must be re-adjusted to compensate.

The 1975–'76 Ford Torino and Mercury Montego steering column is an excellent choice for street rods; the length is about right for 1935-48 Fords and other marques of that era. The best thing about this column is that it has a true universal joint at its base, adequate bearings to support the bottom of the shaft, and the coupler less than six inches from the base of the column.

Forged stock spindles are often usable even after many years of service. The metallurgical qualities of stock steering and suspension components manufactured shortly before World War II are generally high, but there is a caveat. All original steering components must be Magna-Fluxed before installation. This is cheap insurance; a lot of curbs can be sideswiped in 65 plus years!

Most power steering problems make themselves known with a groan or a grinding sound when the wheels are turned. That usually means that a system leak has developed. Sure, you can add fluid to reservoir, but that's not a fix—the power steering system doesn't consume fluid, it loses it.

It may not always be easy for a rodder working out of his home shop, because you have to get the car up on a lift for an adequate visual inspection of the components in the system.

First check the belts and hoses. Look for a loose hose fitting or a belt that is not adjusted to the correct tension. A momentary increase in steering effort can be the result of a belt that is too loose. Be careful not to bear down too much on a belt and burn out a pump bearing.

Next, have a buddy turn the steering wheel from lock to lock, so you can look for leaks. Check all the seals, but pay particular attention to Pitman shaft seals. Carefully check all hoses for chafing and an accumulation of grime and grit—that frequently means the beginning of a leak.

A power steering hose with seepage is trouble. You're dealing with pressures that could exceed 1,000 pounds per square inch (psi), and a mere seep today is likely to be a ruptured hose tomorrow. Not only that, high-pressure power steering hoses deteriorate from the inside out—you won't see an outside problem until the inside is already in bad shape.

Admittedly, it is not always easy to pinpoint the precise origin of a leak. In the popular Saginaw power steering gear, the leak might just be around the sub-shaft seals, but it could just as easily be the rotary valve assembly.

I personally learned the hard way that the stainless steel sheathed hose I was using for the high-pressure power steering plumbing in my '40 coupe was inadequate when it started leaking. George Spillars of Torino Motor Racing (2346 Pacific St., Orange, CA) recommended (and delivered) a 5,000-pound operating pressure, hydraulic hose with appropriate fittings. It's not as flashy as the stainless hose, but it's the right stuff.

Of course, you could have a problem that's not a leak or a loose belt. If the power steering installation is fairly new, check out the alignment of the steering column—make sure the steering column universal joint doesn't bind. If the car feels like it has a poor return to center (as if the power steering gear is locking up) a misalignment between the steering column and steering gear is likely to be the culprit.

Steering column bearings have also been known to seize. If a late-model, tilt steering column has been installed, take a look at the turn-signal switch just for the heck of it—in some models, it can interfere with the steering's return to center. Check the tie rod ends for seizing. Disconnect the Pitman arm and with the wheels off the ground, move the steering linkage by hand.

There are other mechanical problems that are so obvious that they often go unnoticed. One is a loose bolting of the steering gear to the frame. This will cause excessive steering wheel kickback and poor steering control.

Improperly adjusted wheel bearings can also make

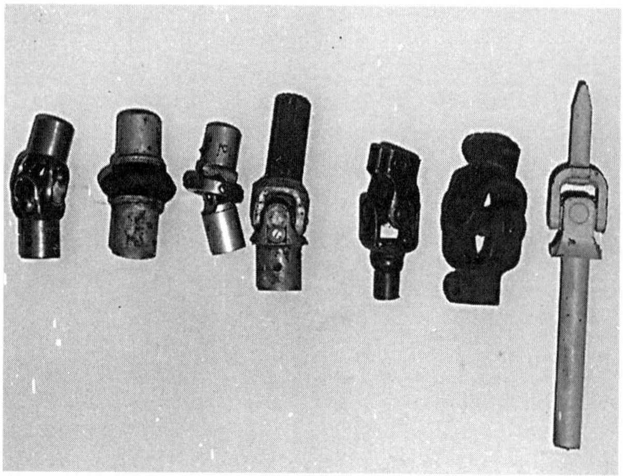

you think the power steering is out of whack. A real mind blower is a pulsing of the steering wheel from left to right when the car is moving along between 5 and 15 mph. The problem may have nothing at all to do with the steering per se. It can be caused by flat spots on the tires, or an area of the tire that does not offer the right amount of resistance to inflation. It is very easy to dismiss the thought of a bad tire when you plunked down $200 for each skin, but if you have a similar problem that plagues you, just borrow a pair of front tires from a buddy to ease your mind. When all the cost-free home inspections fail to yield an answer, it is time to take the car to a shop for a professional opinion.

Universal joints or couplers (as they are sometimes called) suitable for use on street rod steering columns are available from street rod chassis shops. They come in various IDs to slip over the steering column, and depending on the U-joint's particular design, either bolt or weld on. Mil-spec quality components are the strongest and most durable, and naturally they're also the most expensive. Don't forget to have all welded steering joints Magna-Fluxed at a competent shop.

The best Magnaflux facilities are often found near busy airports that cater to folks who build, maintain and fly private aircraft.

Chapter 6

BRAKE TIPS & TECHNIQUES

CHEVY UPGRADES

Here's one for the rare bird that doesn't want disc brakes on his 1940-era Chevy. Check out 1953–'54 Chevy spindles and steering arms. They accept later-style Bendix brakes which offer much better stopping ability than the earlier Huck style. They also provide a five-lug, 4.75-inch lug pattern, and lower the front end 1.5-inches without sacrificing ride.

When planning to install an aftermarket dropped axle designed for 1949–'54 Chevy spindles and a modern disc brake kit in a pre-WWII Chevy, you need to align hone the axle bosses to accept the needed later-model Chevy kingpins. Also, Chevy builders occasionally report a slight wobble with extra-wide tires because the later spindles don't have sufficient kingpin inclination. Keep this in mind when selecting the rubber.

OLDIE, BUT GOODIE

The time-honored brake swap of self-energizing 1953–'56 Ford pickup drum brakes to 1937–'48 Ford passenger cars and light trucks seems archaic these days, but some resto-rod builders still like the looks of drum brakes. The conversion requires National Seal PN 473440 and Koyo roller bearing PN 14116 with only a little grinding on the topside of 1937–'41 backing plates for wheel cylinder clearance. It is not necessary to do any grinding on 1942–'48 models. The race for inner bearing PN 14116 is PN 14276, but you need to compare the inner bearing you removed to the new bearing you are going to install. The older bearing is chamfered where it meets the inside of the spindle. The replacement bearing (PN 14116) is square cut. It is best to have a machine shop match up the parts. The Timken numbers for the outer bearing and race are PN 09067 and PN 09195; they are the same for Bower. Use National Seal PN 47441. The builder is also advised to use riveted linings rather than bonded. Riveted linings are softer and stop better with less pedal pressure.

WHEEL 'EM IN

If you want a front disc brake conversion that doesn't push the wheels outside the confines of narrow stock fenders, first review kits with caliper mounts that bolt to the inside of the spindle.

VERIFY, VERIFY, VERIFY

The rear drum brakes on late 1960s and early 1970s rearend assemblies used on Ford eight-inch axle housings had a variety of brake shoe widths ranging from 1.75-inches to 2.5-inches. An upgrade to a larger size is always a prudent move. However, to avoid buying some oddball mismatch from the local automotive dismantler bring a flange gasket to verify that the bolt spacing on your axle housing is the same.

THE MICO MINI

Builders of rods in the 2,500–3,000-pound range with front discs and drum brakes in the rear who want an under-the-floorboards master cylinder, but find the neighborhood a bit crowded may want to consider the Mico dual-piston master cylinder (PN 02-020-093) originally designed for forklift applications. I used one in my 2,900-pound coupe for years, until I installed a Versailles disc brake rear. Although it has a two-stage feel, it is no bigger than a stock cylinder and mounts to stock bracketry with very little modification. It employs no outside source of power or any trick linkages. Simple hydraulic principles are utilized to develop substantially increased braking power. They are available at truck supply houses.

MUSTANG MC

One of the best manual master cylinders to use on the 1939–'40 Ford with discs-all-around is the one inch bore manual 1967–'72 Mustang. It is easily mounted with an adapter kit from Engineered Components, Inc. (ECI), the premier hot rod brake and bracket outfit (www.ecihotrodbrakes.com). ECI has a variety of modern master cylinder adapter kits in stock. Similar kits for the 1941–'48 Ford, 1937–'39 Chevy and 1940–'54 Chevy are available.

POWER TO MANUAL CONVERSION

Street rodders rarely need to apologize (except, possibly to their better half) for changing something. If you have owned your rod for a long time, you are bound to want to make improvements . . . or simply a change. Many find the vacuum brake booster too big or just simply ugly no matter where it is located.

If you have a car for which there are no cataloged products to make it easy, you must carefully match a new (non-power) master cylinder diameter to the system. However, inasmuch as removing the power-assist booster will increase pedal effort; you must replace the existing master cylinder with one with a smaller bore diameter. (Pressure is inversely proportional to master cylinder bore diameter, so pressure will increase with the same pedal effort.) The trade-off is that with a smaller master cylinder piston diameter, you exchange decreased pedal effort for increased pedal travel. The key is that brake pressure, pedal travel and pedal effort must be carefully matched.

HOSES

High-quality braided stainless brake hoses are far more than cosmetic improvements to your brake system. Because they do not expand as rubber hoses do, pedal firmness and braking force is increased and overall service life is materially increased.

ADAPTATION

When you need to mate AN adapters and conventional automotive hydraulic fittings, use AN-818 compression sleeves and matching AN-819 coupling nuts. Be advised, however, the standard American double flare is 45 degrees and AN fittings are 37 degrees. Don't force them into an unholy marriage. You must use a Rol-Aire 37-degree flaring tool. These are available from a national wide company called Aircraft spruce. (http://www.aircraft-spruce.com/)

PRECISION IS THE BYWORD

Disc brake pads must line up precisely with the rotors. If the pad mating surfaces are not parallel to the rotor, some braking force will be lost. Even if misalignment is only moderate, service life is lessened, but in severe cases, the caliper could suffer damage.

BRAKE VALVES

Pressure valves can be confusing. Most domestic master cylinders for drum brake systems are factory-equipped with a 10-psi residual pressure valve to prevent the brakes from fully releasing and sucking air. When buying a new master cylinder verify whether or not it is so equipped.) Although master cylinders designed for 4 wheel disc brakes ordinarily do not have internal residual pressure valves, a 2 psi valve can be used as an anti-siphon measure.

Proportioning valves reduce the pressure increase to the rear brakes. That is, the pressure to the rear brakes rises at a lower rate than pressure to the front brakes when a specified pressure (called the changeover pressure) is reached.

Pressure-limiting valves are used in some cars with front disc and rear drums, and are similar to a proportioning valve, but not the same. However, the pressure-limiting valve limits pressure to the rear brakes.

Metering valves prevent application of the front disc brakes below a preset pressure in the system.

Pictured below are two Wilwood residual valves. They only look alike on the outside. On the top is a 10 psi valve used for drum brake systems. Below it is a 2 psi valve which can be used as an anti-siphon valve in a disc brake system.

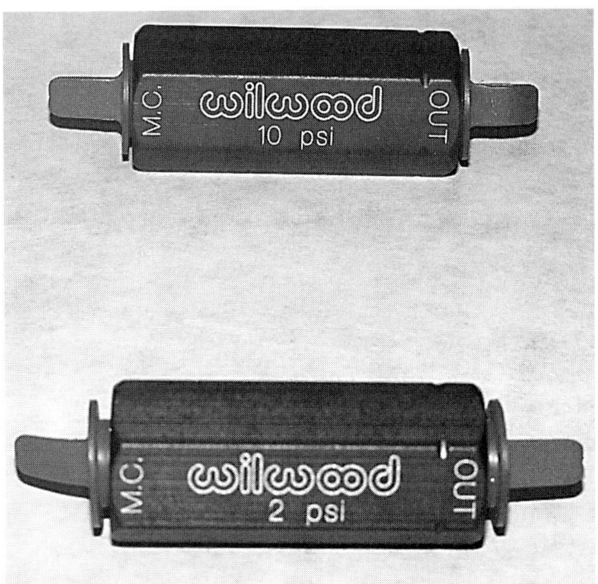

MORSE CABLES

There are lots of cars on the street in the Snow Belt states that have rusted, tight, non-operable parking brakes. This should never happen to a well-maintained street rod. The solution is to replace exposed cables with high quality stainless steel cables housed in plastic sheaths. Boat supply house makes them up to custom order. They are known (generically) as Morse cables.

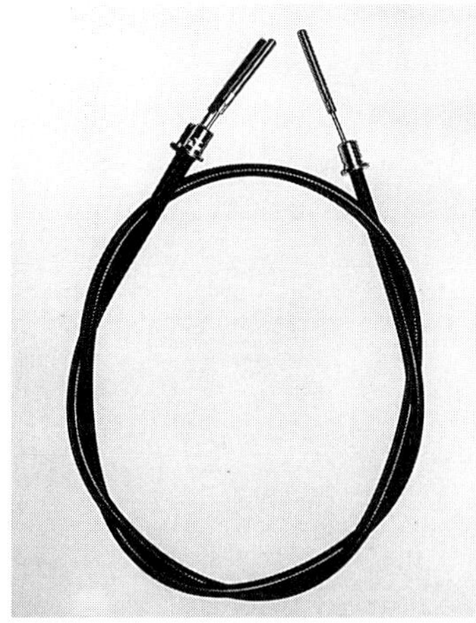

SWITCHES

The lever-type brake light switch used in 1955–'59 Chevy/GMC pickup and similar OEM applications work well in any street rod that has a chassis-mounted master cylinder. The switch is fastened to the underside of the floorboard and is activated by the brake pedal arm. It's simple and foolproof, and available at most auto parts stores.

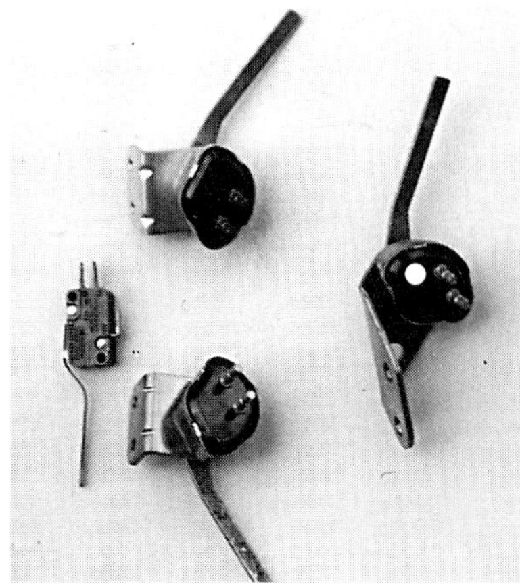

SPONGY BRAKES

Street rods exhibiting recurrent "spongy" brake problems should be checked first for brake lines that are in close proximity to the exhaust system. Barring that, rods with less than airtight brake systems, particularly those that sit for long periods, can develop hydroscopic contamination (absorption of atmospheric moisture) leading to corrosion of system hardware. Double-check all connections, and completely replace the brake fluid from a fresh can. (Saving brake fluid is the ultimate in penny-wise and pound-foolish maintenance.)

A BETTER BLEED

We've all suffered through corralling a buddy to "pump the pedal" when we need to bleed the brakes. And, when the job is finished, there is still that nagging feeling that it is not a "hunert percent." The most efficient method of bleeding brakes is with a pressure bleeder. They are expensive, though. A guy that builds a lot of street rods can probably justify the cost, but all his penny-pinching street rod friends are going to line up to borrow it! It is much more expedient to simply go to a brake shop for the "sleep easy" finishing touch. It's money well spent!

EMPIRICAL BRAKE TESTS

New rotors and brake pads should be broken in. Pick a lightly traveled road and make a number of gentle braking stops to gradually bring the temperature up on the friction surfaces. Let the

brakes cool for a few minutes, and repeat the process twice.

That done, make two emphatic but not "panic" stops. What you want to accomplish is a controlled break-in without overheating the pads or the rotors and potentially damaging both.

Even if you haven't recently installed new rotors and pads, with the multitude of front discs/rear drum and four-wheel disc brake systems, some errors in plumbing or hardware may not show up in ordinary driving. Don't wait for an accident to reveal a problem, make some brake tests on the same clean, dry asphalt road that promises little or no traffic. Gently apply your brakes at 20 miles per hour and come to a dead stop. Get out and look for skid marks. As traffic and continued good handling permits, work your way up to 60 miles per in 10-MPH increments. What you look for (or feel for) is any evidence that the rears and fronts are not coming in at the same time. The potential for unequal line pressures or late lock-ups, front-to-rear is ever present in hybrid street rod brake systems.

WORTH CHECKING OUT

Those of you in Alaska or Arizona (and all points in between) should give Lamb's Lightning Lube a try. It's an ultra-low friction aircraft type grease compatible with disc brake systems. It's reported to maintain a constant coefficient of friction in both very hot and very cold climates. (Lamb Components, Inc., 1259 W. 9th St., Upland, CA 91786-5706, (909) 985-1901. (e-mail: feedback@lambcomponents.com).

IF YOU CAN'T BEAT THE DRUMS

Some street rodders prefer to keep rear axle drum brakes. For those of you who dread going to three different parts stores to gather up all the necessary hardware, ECI offers complete rebuilding kits for 1967–'74 Camaro and 1967–'73 Mustang rears. Everything is included: new drums, new shoes (not relined), new wheel cylinders, and new springs. See photo above.

TECHNIQUE: FLARING AND TUBING

All right, it's time to tackle that plumbing job you've been putting off, and the first thing to consider is what size tubing you are going to use. Old cars came with quarter-inch outside diameter brake line, but 3/16-inch is the domestic favorite today and that's the way modern brake components are designed. Beyond that, even the most inexperienced rodder should be aware that only special steel tubing, designed to withstand high pressure and to resist corrosion should be used. Don't just run out and buy a batch of tubing though, make some preliminary measurements and layout plans.

Use the shortest and most protected route and avoid exposing lines in wheel compartments. (A tire can hurl a sharp rock at incredible speeds.) Don't run your lines too near the exhaust system, and if you pass the tubing through the frame or firewall use a rubber grommet to protect it from chafing.

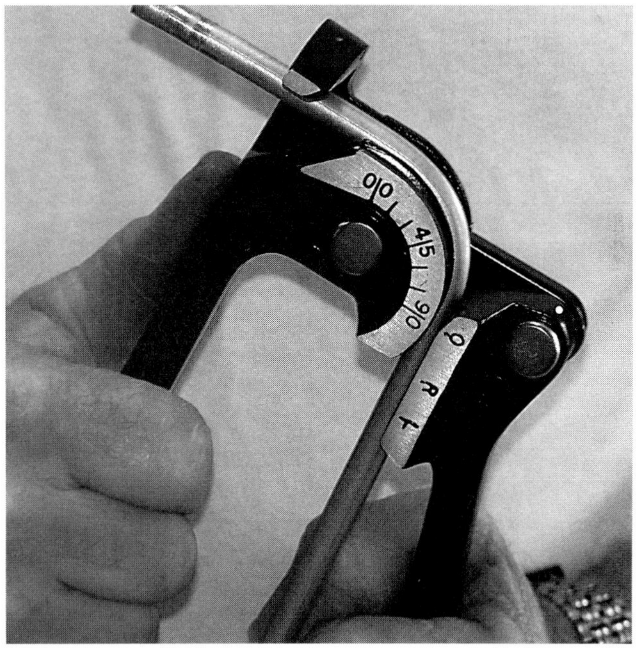

Once you know the approximate lengths of tubing you will need, check with your local parts house for the right size factory flared lengths. Whenever these are available they can save you time. The standard 3/16-inch diameter tubing lengths are 12, 20, 30, 40, 50 and 60-inches. Often you can hit it right on the nose, but not always. Sooner or later you're going to have to hand-make a flare or two, so you might as well learn how.

The most important consideration in making hydraulic brake line is the proper double flaring of the ends of the tubing for the compression couplings. Unless the tubing is properly flared, the couplings will leak and render the best brake system ineffective.

Undoubtedly, you will have to make several tight bends in your tubing, and, it is best to form them before you start flaring. If you are the penurious type, you can use a piece of coiled screen-door spring as long as it is close to the size of the tubing. Twist the spring in the direction that tightens it, thereby reducing the diameter of the coils as you make your bend.

If you are not that much of a penny-pincher, you can buy an inexpensive half circle bender for a couple of bucks that hooks onto the tubing and allows a tight one-inch bend to be formed without collapsing the walls.

However, if you really want to take all of the frustration out of the task, check out one of the more costly professional benders, like the one above.

For more intricate shapes, you will find it to your advantage to make a coat hanger wire pattern of what you need. It's cheaper than tubing and you won't work-harden it or ruin its looks with experimentation.

Before flaring the tubing, cut it to the desired length with a sharp tube cutter. (The blades don't last forever!) Take your time to ensure the tube is cut off square. Ream the inside edges of the tubing with the blade provided on the cutter and chamfer the outer edge with a fine cut mill file. Blow out any metal particles that may have lodged inside the tubing and slip on new compression coupling nuts.

Dip the end of the tubing in some brake fluid; the old timers say this makes a better flare. Clamp the tubing in the correct size hole of the flaring bar and lock the bar in a bench vise. Extend the tubing above the surface of the bar equal to the thickness of the proper double flaring adapter.

(continued on next page)

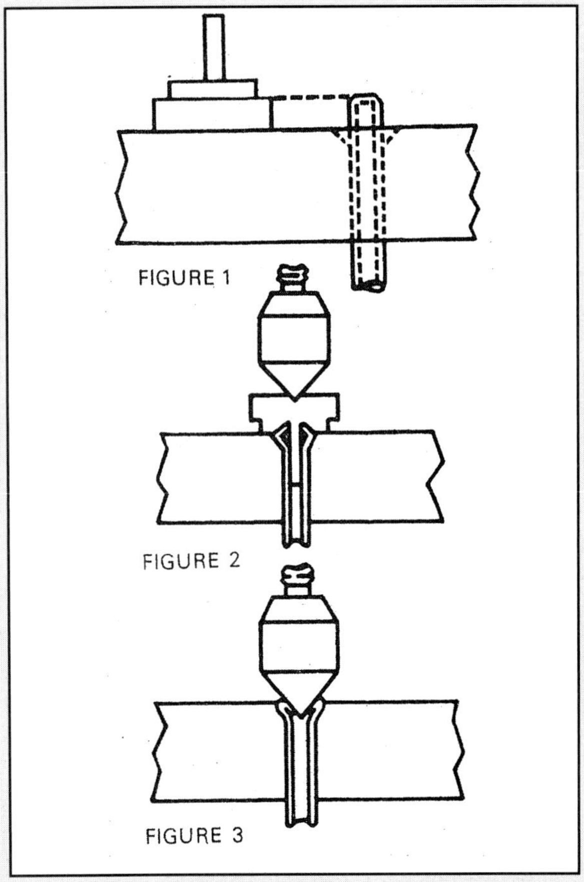

(Fig. 1) Securely tighten the clamping nuts of the flaring bar starting with the one nearest the tubing. When the tubing is firmly gripped it is ready for the first flare forming operation.

Insert the adapter stem into the tubing and place the yoke assembly over the flaring bar with the swivel in the recessed end of the adapter.

(Fig. 2) Gently screw the swivel down until the adapter bottoms out against the flaring bar. This forms the end of the tubing into a bell shape.

Loosen the swivel handle and remove the adapter. Then screw the swivel against the bell-shaped end until the metal folds in on itself.

(Fig. 3) Don't get carried away at this point for if you tighten the swivel too much or too rapidly, you can split the tubing or roll off center.

Before installing the brake lines make certain all fittings and seats are clean. Start each connection by hand to be sure the coupling threads are properly aligned, and then securely tighten each with a tubing wrench to protect the soft corners of the fitting.

Fasten the tubing to the car's frame with the proper size tubing clamps to eliminate the possibility of fractures or fittings loosening and leaking. That's all there is to it. The photo sequence amply illustrates the procedure in more detail.

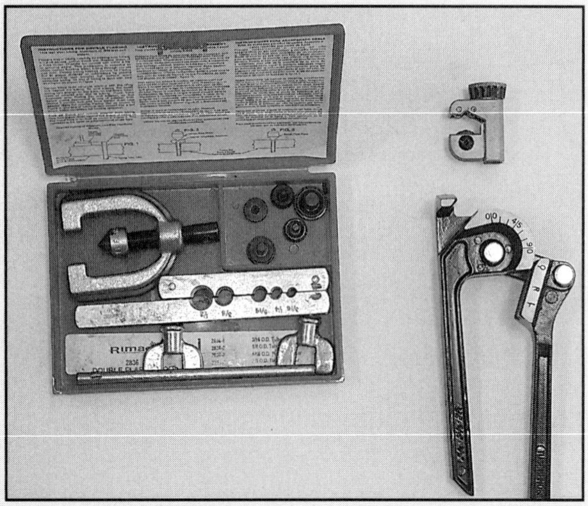

All you need to work with brake lines is a double flaring tool kit, a tubing cutter and a tubing bender.

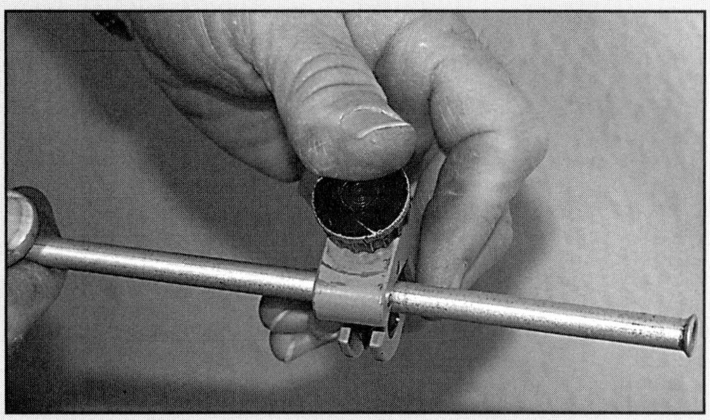

Lock the cutter squarely on the tubing and gently rock it back and forth until a groove is started, then wind the cutter slowly around the circumference of the tubing. Take care to keep your cut as square as possible.

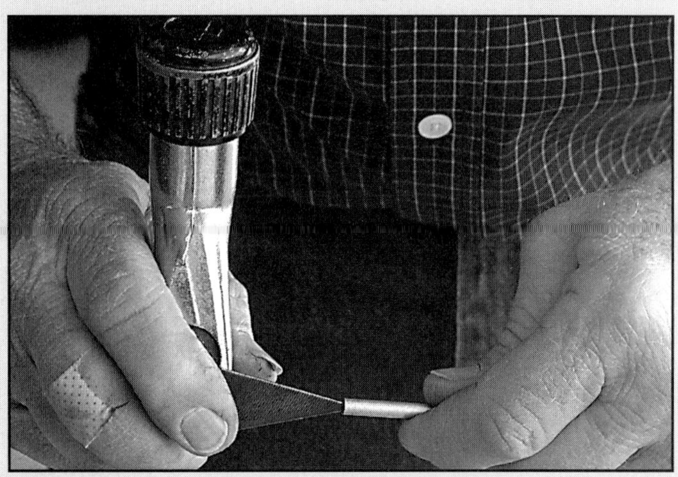

Ream the inner edge of the tubing and chamfer the outer edge with a fine-cut mill file.

condition of the radiator and water passages in the cylinder block. Even a moderate amount of rust is an early warning, and an invitation to have the engine reverse-flushed and the radiator checked by a professional.

A DOWNER

If your existing cooling system is already marginal, you may want to give a second thought to that overbore you were hoping would add some suds to that tired old engine you slated for a rebuild. The combustion heat generated by the extra displacement could put you over the edge.

With the exception of the Ford thin-wall small-block castings that prohibit more than a 0.030-inch overbore, most mainstream street rod engine choices can handle a 0.060-inch punch. But there is a caveat—once cylinder wall thickness is reduced to about 0.080-inch, you can expect overheating to escalate. If you stay within practical limits, however, switching to a radiator cap with a higher pop-off rating will restrain boiling until a higher temperature is reached.

THE DRAIN GAME

There's any number of reasons why you would drain the cooling system. Be that as it may, you want to avoid air pockets when it comes time to put things back together. First, replenish the radiator and motor individually with the coolant of your choice before you replace the thermostat and upper hose.

LOUIE TOLD ME

That done, here's a tip Louie Unser suggested some years ago relative to my Ford-powered Bonneville car. Air pockets sometimes accumulate in the back of the cylinder heads, and just like brake lines, the cooling system has to be bled to get rid of the air. What he suggested was to pull the manifold and note the big coolant passages at the back of each head. Then look at the intake manifold . . . is there a big rusty spot on a corresponding spot of the manifold? Some intake manifold designs block off the rear head passages. If you find a rusty spot, drill and tap the manifold there for a 3/8-inch hose fitting (like those used in fuel blocks) and run an external crossover line between the two heads. Now, the back end of one head can pass coolant to the back end of the other head, just like the internal passages do at the front of the manifold. It's also a convenient place to bleed air out of the system.

SNOW WHITE LTD.

An aftermarket manufacturer of street rod products since 1975, Snow White currently catalogs two cooling system kits that should spark the interest of rodders on both sides of the street. (Chevy and Ford, that is.) One kit raises the Chevy small-block water pump high enough to run a full size fan. The other is a water pump kit that shortens the front of a Ford small-block 302 or 351 (Cleveland or Windsor) by two inches thereby making a FoMoCo engine swap in most any early car or pickup a darn sight easier. (4727 E. Hedges, Fresno, CA 93702, 559/253-1928; email: webmaster@snowwhiteltd.com)

KNUCKLE PROTECTION

Changing a water pump or a fan can be hazardous to both your knuckles and the fragile fins on your radiator! Look through your wood scrap

pile for a piece of eighth-inch plywood large enough to prop up between the radiator and the engine. It will protect both your hands and the radiator should something go awry.

HOSE RELEASE

Whatever goes on a street rod has to come off sooner or later. A silicone lubricant such as WD-40 sprayed inside the ends of the hoses prior to installation will prevent them from sticking to the radiator, water pump necks or heater hose nipples.

HOSE REMOVER

If you didn't take the above precaution, and those hoses seem to be welded to their respective

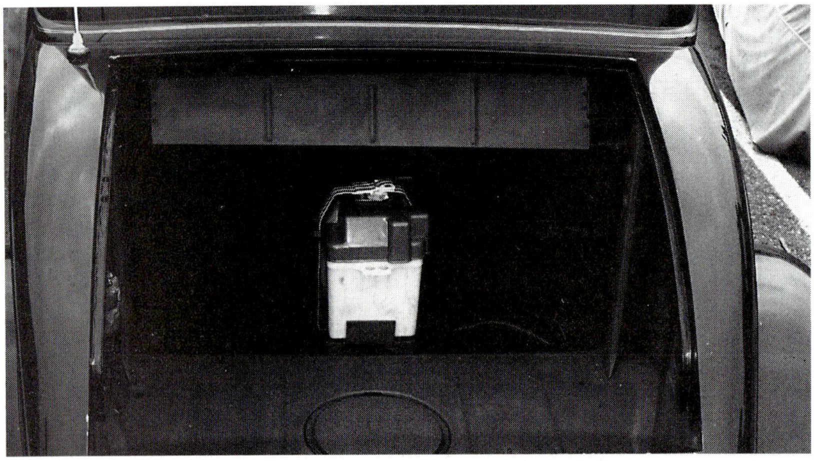

HIDE & SEEK

Battery placement has been a problem for as long as I can remember. Street rodders have put them back under the hood where they were originally, in the trunk, under the seat, under the floorboards, and even in the wheelwell. When they first came out, maintenance-free batteries made it less troublesome, and nowadays, the new, compact, high-amp batteries have made it even easier.

MORE POWER TO YOU

To get an idea of the minimum cranking amperage you need when selecting a battery; multiply the engine displacement by 1.5. It isn't prudent to buy one with fewer cranking amps.

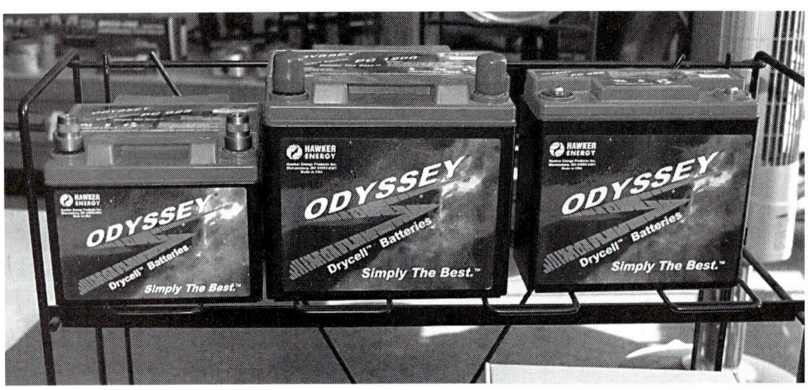

PROPER VENTILATION

With the variety of high-quality batteries available today, it is also well to verify whether or not your new purchase will or will not emit gas. Since street rodders typically like to hide the battery, it is mandatory to accommodate those that do emit vapors. A fully enclosed marine-type battery box is recommended, and depending on where it is mounted, modify the box for a pair of half-inch ID hoses that will vent the gas to the atmosphere.

BIGGER IS BETTER

If you do put your battery in the trunk, remember the cable to the starter should be as stout as practical. I've used half-inch thick copper arc welding cable for years with great success.

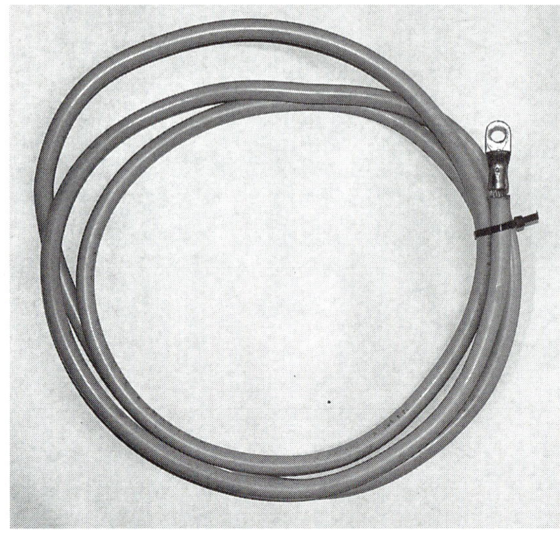

CABLE PROTECTION

Another thing, when running the positive cable from the trunk to the engine hold, make absolutely sure it is protected from any damage to the insulation that could result in a short circuit.

It isn't necessary to use a long negative cable inasmuch as the chassis serves this purpose. You must, however, run an equally stout ground cable from the engine block to the chassis.

CRIMP IT RIGHT

Many solderless electrical terminals sold by auto parts houses have a seamed barrel. Pay attention to it when you crimp the terminal to the wire. The seam should be opposite the male part of the

crimper die. This prevents accidentally opening the seam. This is not a problem with extruded, seamless terminals. Nevertheless to avoid falling into bad habits, position the crimper on the terminal so that the side opposite the ring is also opposite the male die.

STOP THIEF!

If you are going to install a "disconnect" anti-theft switch, do so in the negative (grounded) side of the electrical system. This prevents a short if there is a problem at a later date.

AND THE BEAT GOES ON

If you wire your electric fan into the ignition circuit, when you shut the engine down, the still freewheeling fan motor will act as a generator for a short time, feeding current back into the ignition. The remedy is to split the circuit, easy to do if the ignition switch has two poles. If not, run a separate switch for the electric fan and turn it on only when you need it.

KEEP COOL

Excessive heat from the exhaust system can cause the starter to crank slowly and even fail prematurely. A one-inch minimum air gap is needed to prevent the development of electrical resistance in the starter. To be double safe, fabricate a heat shield out of a piece of stainless steel. If space is at a premium, and the stock starter sometimes drags, you can solve both problems with a high torque starter that is smaller in girth.

OLD WIRING

If the wiring in your street rod was done before you acquired the car (or you did it, and you just can't remember), check the main charging wire that transmits current from the alternator to the battery. If it is only 12-gauge, replace it with 8-gauge wire. The heavier wire easily handles up to 120 amps of current.

You can still protect the charging circuit with a 12-gauge fusible link, a piece of wire that acts like a slow fuse. If there is a system malfunction, the fusible link blows first (and safely).

NEW WIRING

It only looks easier to wire a rod when the body is off. In truth, you know where your obstacles are when all the various components are in place. And, when you begin, wire the motor first.

CLAMPS AND LOOMS

No matter how neatly you bundle topside engine wiring with neat little black or colored wire-ties, the wiring insulation becomes brittle over time. A wire loom made of painted or plated tubing tastefully curved around engine components and fastened with small aluminum or PVC Adel clamps can protect the wiring and reduce topside clutter, but it takes some planning.

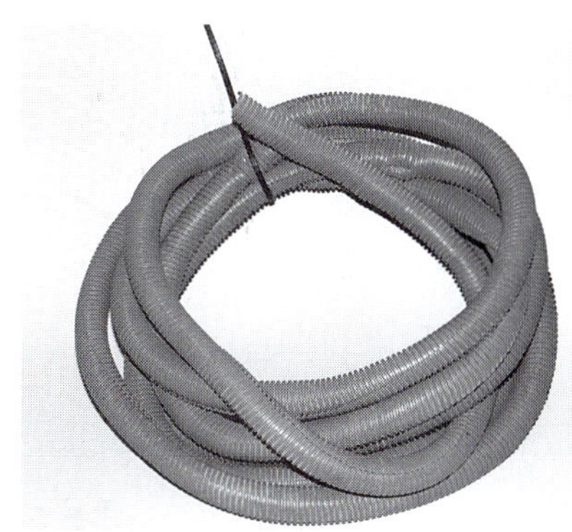

If you need to bundle several wires, you may want to consider color-coordinated split flexible tubing even though this tubing can quickly become gaudy. Fortunately, most of us have a better half who can double as an interior decorator. Ask your wife's opinion!

IT'S A WRAP

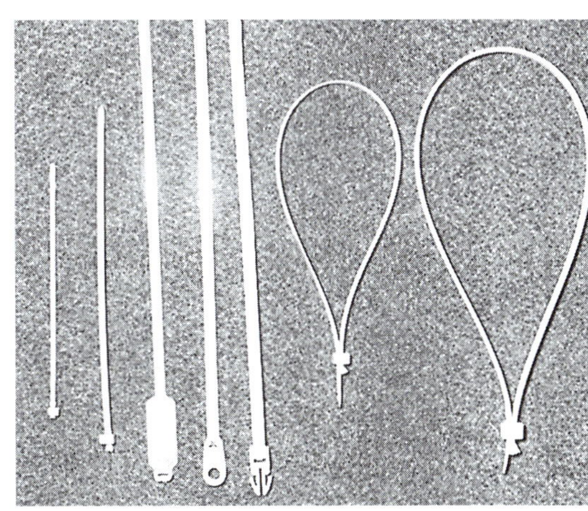

Plastic tie wraps are used throughout when wiring a street rod, but don't overuse them. One about every six inches is plenty. They come is various configurations, sizes and colors. Their only drawback is razor-sharp cut-off ends when the tail is clipped with side-cutter pliers.

FOR THE TOOL COLLECTOR

There's a tool for just about everything. If you want a super neat excision, or just don't like pulling the tail of a wire tie by hand, look for one of these at the electronics store.

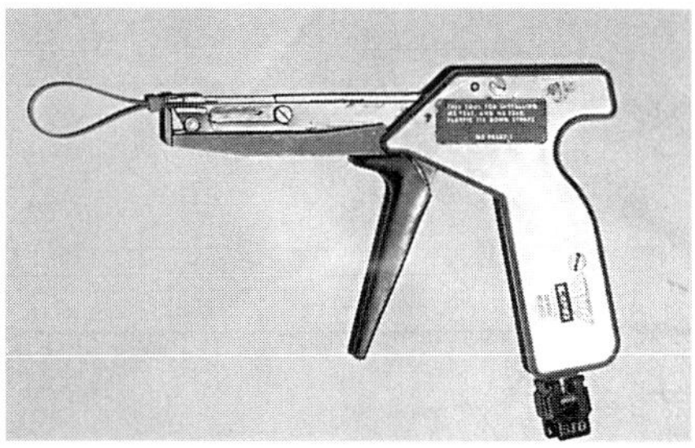

Otherwise, careful use of your soldering gun can quickly blunt those dangerous little blood-letters.

READ THE REGS

Make sure you're on safe ground with regard to your state's headlight and taillight regulations. That highway patrolman may not agree that your resto-rodded 1940 Ford Standard coupe with its single taillight is no problem when you're doing 50 in a 45 on a dark rainy night.

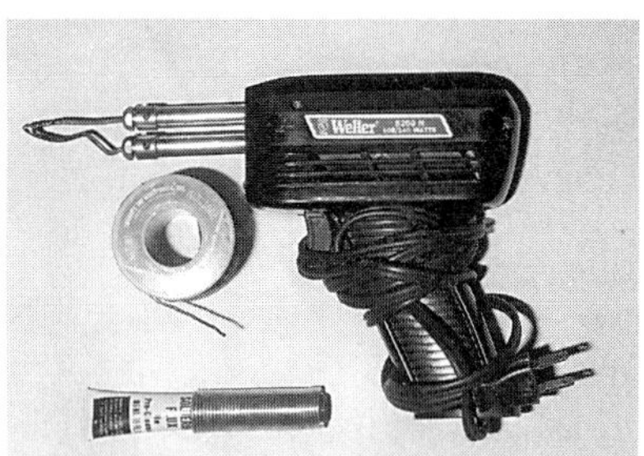

GET IT RIGHT!

Of course you know that the power lead to the coil goes to the positive post (in a negatively grounded circuit) and the lead from the distributor goes to the negative post. If they got mixed up while you were rearranging things, however, the engine may fire up and run, but coil output will be a shadow of its former self.

Another thing, you can't afford to be cavalier about polarity. If you get your jumper cables reversed, the diodes in the alternator are history in the blink of a careless eye!

DASH DOINGS

If you are still abuildin' or you have your dashboard out for modifications, take the opportunity to paint the inside gloss white. You'll be amazed how much better you can see the wiring the next time you have to go under with a droplight or flashlight.

Speaking of droplights, I've been burned (literally) too often to ever trust my hands (or the coupe's carpet) around a conventional incandescent

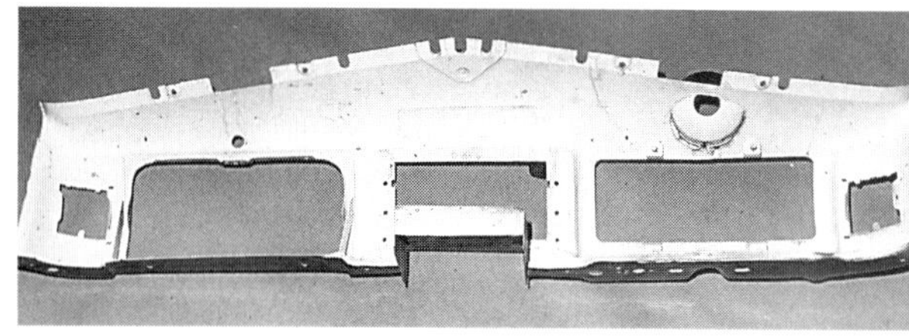

bulb type again. I've had this cool (no pun intended) fluorescent model for so long I don't remember where I bought it. I'll bet you can find one like it, however.

ITCH TO SWITCH

Want to get turned on? Check with all the potential resources, not just the local auto parts store. You'll be surprised at the variety available in truck, marine, electronic, and RV parts houses! One or more of them is bound to have a switch that will function properly and look right in your street rod.

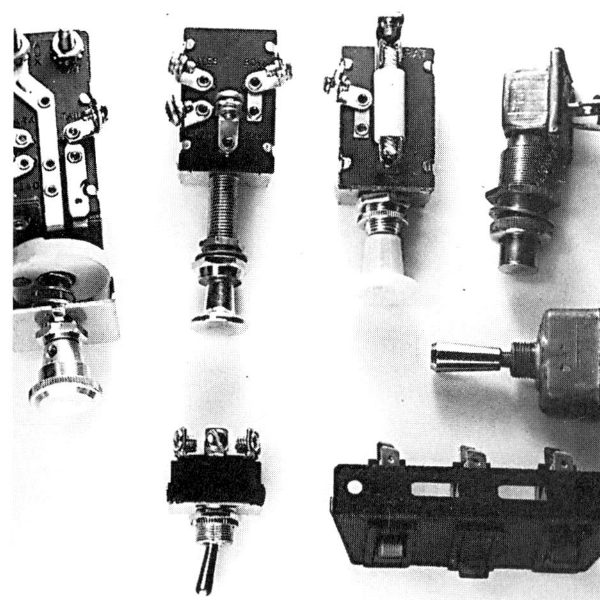

NOTHING BUT THE BEST

Even if your wiring tasks are only periodic, you want the best crimping tool available. Specialty electronics supply houses usually have the cream of the crop. The better of these two (in terms of quality and service life) should be obvious.

Even though you are likely to use solderless terminals, always keep the soldering gun close at hand. You will need it from time-to-time. By the way, don't forget, solder with an activated rosin core is the industry standard for most electrical and electronic hand soldering.

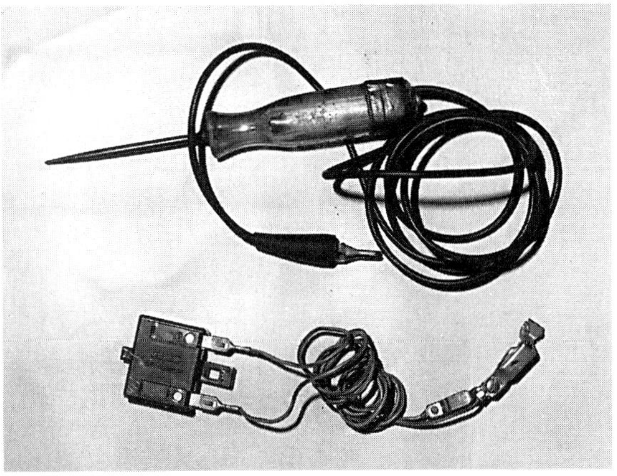

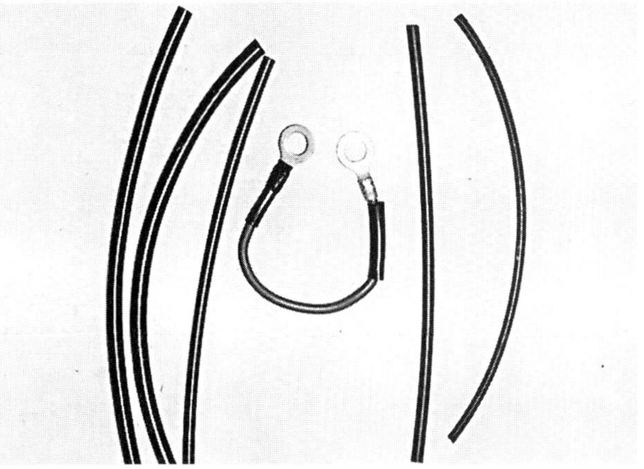

CHECK IT OUT

Test lights, either commercial or homemade out of a "key-in-the-ignition" type buzzer are invaluable for checking your circuits.

MO' BETTER

Don't use adhesive-backed electrical tape if you can avoid it. Heat-shrink plastic tubing, available in several diameters, makes the wiring task neat and easy. I've already mentioned Adel clamps offer a quick and attractive way to bundle and secure wire looms. And, that they are available in padded steel as well as molded plastic. Here, I want to emphasize the availability of a variety of inside diameters. Neatness compels you to match the ID to bundle girth.

Chapter 9
BODY/PAINT TIPS & TECHNIQUES

A NEW PROJECT

The place to begin reworking an original car into a street rod is the body shell. The restoration of major sheet metal is costly, so you must be sure that the body is worth the subsequent time and money typical street rod modifications require.

ENGINE SWAPS

There are two major determining factors in any potential engine swap: Size (which naturally includes weight) and configuration of the intended engine. Obviously, size and configuration are directly related to engineering effort, but don't forget eye appeal. (If you question the latter, just think back to engine swaps you've seen that were obviously crowded and just didn't look right. When you stray, you're often on your own. If at all possible, find a "similarly situated" engine swap at a car show or rod run for evaluation before you start moving metal around. An engine installation shrouded by the fenders and firewall can be deceptive, but even a rough idea is better than an expensive shot in the dark.

Pan Problems—It is useful to know the location of the oil pan sump (fore/aft/middle) of the engine originally in the vehicle even though you are (hopefully) planning on an updated steering and front suspension.

Specific Dimensions—It also helps to know the length and width of the original engine from the water pump to the transmission interface unless you are already planning on custom formed sheet metal or fiberglass.

Research—Where are you going to find these admittedly obscure dimensions? The best single resource I can recommend is *Hemming's Motor News*. Not only are restoration clubs listed, the sellers of original parts will almost always respond to a short letter that includes an addressed, stamped envelope.

Compare—The dimensions and configuration of the original engine and transmission coming out of the vehicle to those of the power package going in. The engine hold has to be large enough to accommodate all dimensions, but most pre-WWII passenger car and pickup holds are smallish.

The point is, you want to know if and how much the firewall and fender wells have to be reshaped, and if the floorboard (as well as the chassis) has to be modified. That way, the labor costs and extended time requirements can be realistically factored into your project budget. Any doubts as to why small-blocks are typically favored over big-blocks?

CHOPPING AND CHANNELING LIMITATIONS

Hammered Model A street rods look great, but man, you'll want to drive the thing sooner or later! I recommend a conservative approach with regard to major body modifications. Don't drop the body lower than the bottom edge of the chassis or take more than three inches out of the top.

FIBERGLASS PAINT STRIPPING

Sometimes a new paint job isn't dictated by the condition of the existing finish. Sometimes the street rodder simply wants a color change. Furthermore, it's been my observation that this more frequently occurs in the roadster ranks than in four door sedans. And, I think it is a safe bet that there are more fiberglass roadster bodies than steel being eyed by their owners. The very fact that they are small also encourages the owner to pocket the cost of professional stripping whether or not he is going to spring for a professional paint job.

Well, stripping fiberglass of its paint isn't all that different from stripping sheet metal. However, the very nature of the material requires a bit more care. Before you don your gloves, long sleeve shirt and protective goggles consider this . . .

Cover no more than a two square foot area with a heavy layer of paint remover. The remover should be applied with a brush, and should be stroked in only one direction, not back and forth.

If you choose not to remove the doors, all of the crevices should be masked off with half-inch wide strips of masking tape to prevent paint remover from dripping in. Once the body has been stripped, the paint can be carefully removed from the door jambs.

Depending on the paint and the ambient temperature, the remover will take from 10 to 30 minutes to soften the finish, and will bubble when it is ready to be scrapped off. Higher temperatures will speed up the process.

Nevertheless, the paint remover should not be left on the body or component for more than thirty minutes as it will soften the fiberglass. The bubbled paint should be scraped off with a Teflon body putty spreader to prevent scratching the 'glass. It takes two or three applications of stripper to completely remove the paint, but each two square foot area should be completely rinsed with water after it has been stripped.

It is best to use the stripper only until you get down to the primer. Sand the primer off with 180-grit wet or dry paper and water. A good quality, dense sponge makes a perfect sanding block.

After sanding is finished, the body should be cleaned with a wax and silicone remover. Before you call it quits, though, make sure you've removed all traces of the old paint; otherwise the new finish will discolor or even crack in spots. It must be emphasized that paint stripping a fiberglass body by any mechanical means is fraught with danger, and even hand sanding can have its downside if one gets careless.

If, after reading the above, you decide that you really don't want to do the job yourself, the Redi-Strip Company has also perfected a technique for fiberglass bodies. The process makes use of powdered baking soda, and is pressure applied in a manner similar to the application of powdered walnut shells and so-called plastic media. However it is less aggressive than these methods and does not cut into the gel coat. For more info on Redi-Strip, see page 83.

WHERE THE RUBBER MEETS THE DOOR

Old cars leak wind and rain badly enough without reducing the efficiency of the reproduction door seals. Often the builder has installed the rubber without carefully reading the directions. For instance, the upper door rubber seal on 1937–'48 Ford closed cars must be cemented to the body, and spans from the upper door hinge to the back corner of the doorpost. The lip of the seal is at the bottom.

All too often, the seals are mistakenly glued to the door, an error that likely comes from the location of the rubber hinge seal on 1935–'48 Fords. Here, the rubber is cemented to the door, from the top of the upper hinge almost to the bottom.

DAZED AND CONFUSED

Many street rodders are confounded by the similarity of front fenders and hoods from the 1939/'40/'41 Ford passenger cars and pickup series.

First, passenger car hoods and fenders are not interchangeable with the look-alike pickup sheet metal. Here's what will interchange without major metal work: 1939 DeLuxe and all 1940 passenger car front fenders.

Although the respective chrome ornamentation and stainless steel trim differ, the 1939 Ford DeLuxe hood and grille (as a unit) will interchange with the 1940 Standard hood and grille (as a unit) and the 1940 Deluxe hood and grille (as a unit).

Likewise, 1940 and 1941 Ford pickup front fenders and hoods are interchangeable, although there are minor chrome ornamentation and stainless steel trim differences between hoods. But because the 1941 models offered six-cylinder and four-cylinder motor options, hood-latching hardware is different.

SPRAY-ON PROTECTION

Thomas Bank of Pennsylvania advised us awhile back that the popular spray-on pickup truck bed liner can be used to protect the underside of street rod fenders from the road shrapnel that causes star bursts in the topside paint. It comes in different colors and can be applied in different thicknesses. Depending on the facility's willingness, you may be able to bring loose fenders in for the treatment.

SHAKE, RATTLE AND PRESS

Such resources may not be available in your area, but an auto parts store is sure to be within driving distance. Swing by and order a few aerosol cans of 3M's Body Shutz. It too is a great starburst protective and sound proofer. You'll have to apply it yourself, however.

If your car has been on the road for some time, but you're not in the mood to remove the fenders, you'll have to clean up their underside. Head over to the quarter car wash and spend a few minutes with each wheelwell and hot, soapy, high-pressure water. Back home, pull the wheels and finish the clean-up job. After drying the sheet metal, tape off the bolt heads and washers, headlight/taillight wiring, and the outer edges of the fender. It's not as messy as it sounds, but eye protection and old clothing is a must.

WANTED: SPANDEX PAINT

Everybody admires the results of a custom paint job. But custom paint jobs have one significant drawback—fragility. Two-stage applications are those with a base coat, some containing very fine metallic particles. The topcoat is a layer of clear paint, and it is particularly vulnerable. As the paint cures, it hardens with very little elasticity. In time, it can become so hard and brittle that it won't stretch as the body metal expands underneath it.

Cracks in the paint usually start on the roof, hood or trunk lid. There's only one prevention—minimizing exposure to significant changes in temperature. A garage is best, of course, but not always available, especially on long rod runs. A top-grade, soft flannel-lined car cover is an absolute must, and should be carried in the car for availability as needs arise. Otherwise . . .

. . . once cracking has started, the only way to fix it is to sand the car down to the primer coat, re-primer and repaint the car. Sanding off the clear layer and re-spraying will only result in an uneven-looking finish. Sorry 'bout that.

YOU SHOULD BE SO LUCKY

If you are lucky, and only the roof, deck or hood is cracked, and the paint scheme permits, you can repaint just that section. Nevertheless, inasmuch as the remainder of the body already has all the exposure it can handle, you'll have to be doubly careful it isn't further subjected to the elements.

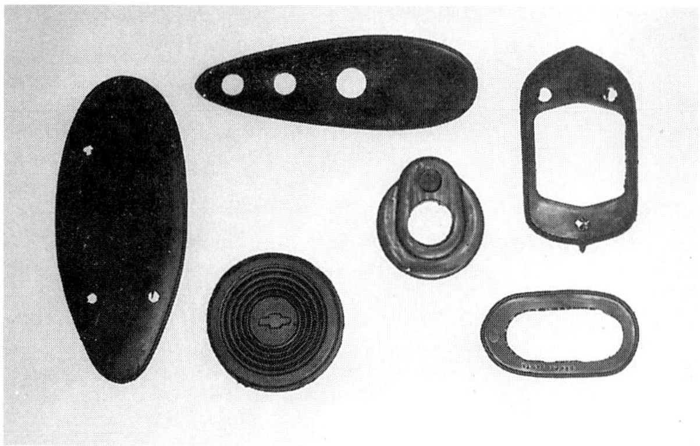

WHEN IT POURS

Rain leakage is a major problem in rods with swing-out windshields, and, unfortunately, is sometimes incurable. That's why many street rodders eventually want a fixed windshield. Chevy S-10 rubber molding in the hands of a competent glass shop technician works very well.

BOW-TIE REPRO

Looking for reproduction rubber products for the 1929 through 1954 Chevrolet? Check with I & I Reproduction, Inc. E-mail: inquire@iandi reproduction.com, or contact them at 15513 Vermont Ave., Paramount, CA 90723, (562) 531-8117.

SAFETY CHOICES

I've given much thought to street rod safety since my coupe was rear-ended in the 1960s. In my opinion, the body's construction material (fiberglass, composite or steel) is not as important as the configuration of the car.

Let's consider open cars first. Roadsters are flimsy, and I seriously doubt that 'glass, composite or steel would make much difference if the car were "T-boned" at any speed. However, a discrete, padded, two-part roll "cage" (one part in the firewall area, and the other just behind the seat) would make a difference. An unobtrusive roll bar that comes approximately to the height of the driver's head looks "right," and along with properly installed seatbelts, provides a significant measure of protection. Convertibles, being larger and more massive, do offer more protection for driver and passengers. Nevertheless, a pair of roll bars would be desirable in any serious accident.

With coupes and sedans, sheet metal has an advantage over fiberglass, but again, design is the prime factor. I do feel that my own coupe (sans the roll bar I installed in the 1970s to run Bonneville) is reasonably safe with only lap belts. The architecture of the stock turret (that area above the stainless trim line) forms an integral "cage." By the way, my kids, who were sitting on the "opera seats" at the time of my 1969 accident, weren't scratched, although the rear end impact tore the engine and transmission from their respective mounts.

STREET LEGAL FENDERS

Generic trailer fenders are not a bad idea for street roadster rear tire covers when such are required. They need a bit of massaging to make them look neat, but they are much less expensive than repro (much less originals!) and are strong.

Motorcycle fenders were very popular for front wheels in the old days. The rod builder went down to the local Harley or Indian shop where they had several styles of steel fenders and miscellaneous pieces of bracketry. With the variety of motorcycles available today, there are even more options.

Mounting motorcycle fenders on an exposed front end suspension with lumpy disc brake assembly is more difficult than it was with yesterday's flat, drum brakes. Nevertheless, you should be able to find some place upon which to weld tabs and mount the brackets. Creative experimental fabrication is called for, as well as trial and error fitting to come up with a workable solution. Before you head out for a test drive, jack the wheel assembly up and down and the turn the wheels lock-to-lock to simulate suspension movement and confirm adequate clearance.

BAD VIBES

Mysterious vibrations plague even the finest of street rods. Isolating their root cause is an exercise in patience at best, and a world of frustration at worst. Thankfully, it is usually not so technically involved that a practicing street rodder can't diagnose most of them himself, especially if he has some kind of format to follow.

To determine the origin of noise and vibration, you must duplicate exactly what is bugging you. Then you have to evaluate the problem under changing conditions. That's the only way you can work your way through the process of elimination needed for a satisfactory repair.

Find an open, safe area for your test drives. Listen closely to the vibration or noise while you slowly accelerate; then coast down in neutral. Try downshifting, using engine friction and compression for braking. Run the engine up while in neutral, but keep your hand on the shifter.

Make a few fast, than slow, brake-applied stops. While at a stop, stand on the brakes and slowly run the engine up while you listen for vibrations or shudder.

Roll out again and steer gently left and right. If road and traffic conditions permit, weave a little.

(Don't overdo it, however!) Try a couple of slow and fast standing starts. What does the vibration feel like? What does it sound like? What does it respond to?

RATTY RAIN GUTTERS

Replacing rusted out rain gutters on most early cars is one tough (and expensive) job. If you look closely, you'll see the sheet metal forming the rain gutter tucks under the metal forming the roof of the car. When the gutters are too far gone, trim off what's left and smooth out the rough edges.

Many body styles, however, then seem to have too much "open space." Solution? Pinstriping or a trick paint scheme often pleases the eye.

BED WOOD

If you're not particularly handy with woodworking tools, Bruce Horkey's Cabinetry (46284 440th St., Windom, MN 56101; Phone: 507-831-5625), offers pickup beds and replacement wood floors for a wide variety of Ford, Chevy/GMC and Dodge pickups.

KEEP IT PRISTINE

You keep your baby garaged whenever you're home and you never leave home on a long rod run without your double-throw-me-down felt-lined custom-made car cover. But no matter how cautious you are, Murphy's judicial envoys will find you! Life is hazardous for cars as well as people—tree sap, industrial fall-out, and our feathered friends never heard of waste management! Clay Magic is a product we used for years at the body shop to eliminate the unavoidable paint overspray. Keep a can on the shelf in your home garage.

TECHNIQUE: REDI-STRIP GUARANTEED RUST REMOVAL

The Redi-Strip Company of Roselle, Illinois, bills their service as "fast, sure, safe rust removal from anything." Nothing better dramatizes that claim than these shots of a 1939 Ford convertible slowly being immersed, wheels, tires and all, in a proprietary alkaline electrolytic solution. And then, being flipped like a pancake to ensure 100 percent oxidization eradication.

The chemist classifies it as ferric oxide, the layman knows it as rust, but regardless of the foul names you may call it, it always adds up to bad news for the street rodder. Rust has caused more backyard wrench twisters to forsake a rod project than an irate wife. 'Cause in the cold light of day, it takes but one long look at the rusty hulk that was once a passenger car to make you want to toss in the grease rag. It simply goes with the game—cars that were semi-abandoned years ago are likely have faced the worst Mother Nature can deal out.

Coupes and sedans suffer enough, but open cars are by far the most vulnerable to the ravages of the elements. The floor pan and other interior metal of a topless car have often rotted away, adding insult to injury when the time comes to take on an exercise in restoration. Just the paraphernalia necessary for amateur rust eradication is forbidding—paint remover, muriatic acid, wire brushes and wheels, scrapers, open-coat/closed-coat/fine/medium/coarse discs and sandpaper, goggles, gloves, dust mask and one and on, ad nauseam. And don't forget the need for a generous helping of time and muscle. Of course, should you weaken during the attack, either from fatigue or dermatitis, the local sandman will gleefully do his destructing thing to your precious sheet metal.

But wait, don't trailer your car or accumulation of parts to the blaster's dungeon—there's a better way. A way so versatile, you won't believe it until you see the results for yourself. In the early 1960s while seeking a better paint-removal process, the folks at the Redi-Strip Company accidentally discovered a method that not only removed paint . . . it removed rust from steel without etching or embrittlement. Technically, it is known as the alkaline electrolytic immersion process. A mouthful to be sure, but whatever you call it, don't call it "acid" dipping. That's a real four-letter word at the Redi-Strip facility. John Carlyle, the proprietor, doesn't want his process to be confused with "pickling" or acid stripping because of the latter's inherent threat of hydrogen embrittlement.

The Redi-Strip tanks are filled with an electrolytic alkaline solution heated from 140 to 160 degrees during the derusting or paint stripping operation. An imposing generator mounted near each tank drives an electric current through the solution in a manner not unlike electroplating; only the result is just the opposite. The dipped part acts as the cathode (the negative electrode) and rust molecules are "dissociated" from the work piece. Although "hot tanking" or caustic soda solutions have been available for decades, the patented Redi-Strip rust removal process is unique.

The process easily removes all coating up to and including epoxy. In fact, the proprietor flatly states that most anything applied to a steel surface can be removed. Best of all, it only removes the bad stuff—rust, grime, undercoating, ancient paint and unwanted hair. Even plastic body filler is unharmed as long as it was originally applied properly.

Moreover, the process does not affect dissimilar metals. Because heat is required to speed up the chemical activation, the tanks are thermostatically controlled for various types of metal or bimetallic items. Aluminum, as always, is a problem and will disintegrate in a solution scheduled for steel. However, when steel and aluminum are conjoined and separation is not practical, the temperature and formula are adjusted to meet the specific stripping requirement of the work piece. Although the process doesn't affect glass or solid wood (except to slightly raise the grain) the lamination in automotive glass will soften, and the glue in plywood will dissolve.

Let me point out that I am talking about two independent processes—paint removal and derusting. A non-rusted part can be stripped of paint and grease without being placed in the derusting tank. Or, if the part has partially rusted, but for some reason you do not wish to remove the coating, it can be stripped of rust without damage to the finish. For example, a manufacturer of electrical components had a sizable batch that were corroded and oxidized. Before tossing them out they decided to give Redi-Strip a shot at restoration. When all was said and done, a critical analysis of the components found them to be fully functional.

More pertinent to street rodders, especially those who are resto-building cars for which there are no reproduction parts, is the all-too-common problem of pitting in die-cast grilles and ornamentation. They may look great when they first come back from the plater (assuming they haven't been overly polished) but within a few months, the chrome starts to blister! The reason is that the pickling and polishing the chrome shop does simply will not always stop the insidious pitting. This can be prevented by the Redi-Strip process. Pitting is an oxidation akin to rust and the electrolytic action actually pulls the contaminating molecules right out of the "pores" of the die cast alloy.

To properly prepare a die cast item for a long-lasting plated finish, first take the piece to your favorite chrome shop and have them strip away the thin layers of chromium, nickel and copper down to the base metal. Then send the part to Redi-Strip for derusting immersion. When you get it back, bring it to your plater for a heavy copper coating and re-chroming. This regimen is somewhat involved to be sure, but if those rare Whippet windshield wiper towers can be saved, isn't it worth going to the extra trouble and moderate additional cost?

When the work piece is removed from the derusting tank an alkaline residue forms. The residue acts as a temporary surface rust inhibitor as long as the part is kept dry. However, moisture from external sources will attack the coating. That's why it is always best to have sheet metal "phosphated" at Redi-Strip. When you get it home spray it with a high quality primer-surfacer/sealer.

If throughout this dissertation I have sounded impressed with the alkaline electrolytic method and its endless list of opportunities for the street rodder, I am. Yet, because Redi-Strip is the only company possessed of the formula, and it is located in Roselle, Illinois, you may feel that the foregoing is only of academic value unless you live within a reasonable driving distance of metropolitan Chicago. But not necessarily—just find a buddy or two in the same dilemma as you, pool your rusty parts and ship them by truck to the Redi-Strip Company, 100 West Central Avenue, Roselle, Illinois, 60172. Naturally the details must first be ironed out by mail or a telephone call to John Carlisle (630-529-2442). This practice has proved quite efficient for a number of street rod clubs.

In short, if you can get your rusty sheet metal and pitted die cast ornamentation to Redi-Strip, you'll be as amazed with the results as I have been, time and time again.

There's simply no way to completely eradicate rust in such complex areas as an open car's door jamb and top frame.

Even relatively exposed areas such as wheelwells could never be sponged free of rust with hazardous muriatic acid, or heavens forbid, sand blasting.

As impressive as it may be to watch an entire car body sink into the rust removal solution, de-rusting more prosaic parts and die cast ornamentation is just as critical for the completion of a real steel street rod project.

TECHNIQUES: PREPPING A MODIFIED CHASSIS FOR PAINT
STAGE ONE: THE METAL WORK

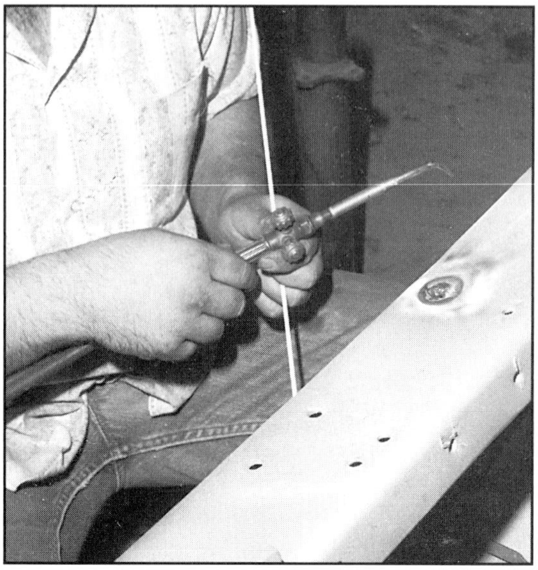

Like every other part of a well-built, well-detailed street rod, an original chassis that has been rehabilitated and looks like a million bucks probably involves almost that much professional labor. Fortunately for the rod builder, this is one area of metal work where extensive experience isn't required. What is needed is patience and access to an oxy-acetylene set and a body sander. Even if these tools are not in your home shop, they can be rented at nominal cost. And since welding supply stores offer dozens of how-to pamphlets on basic brazing and leading, I won't parrot them here.

The real trick is acquiring a bit of skill with that big heavy disc sander. But rest easy, it takes surprisingly little time. In going over the steps required to prepare an original, but upgraded frame for paint, it will be assumed that it has been sandblasted, thoroughly inspected for cracks, and all necessary major repairs and modifications completed. These tasks are likely to have been farmed out to the pros, but the remaining labor is well within the amateur 's capability.

Several varieties of sander/grinders—oscillating, belt and rotary—are used in auto metal finishing. Of these, the 7-inch diameter disc sander is often favored for frame work. A single speed, 4,600 rpm, one-horse model is economically priced and readily available. If you are more than an easy-chair street rodder, it will frequently find its way into your hands. Learning to use it properly, however, will take a bit of practice. Before you hit the disc sander's "On" switch, be sure to don your safety goggles or face mask.

Grasp the sander firmly, but be aware that heavy pressure will slow the cutting action and prematurely wear out the abrasive disc. Depending upon the size of the area being worked, the sander should be used with a sweeping, back and forth motion. This provides the smoothest continuous coverage. Avoid using a circular motion, and do not hold the sander in one spot, or you will burn the metal. (This may not so critical on a frame, but it develops bad habits, and sooner or later you'll work on lighter gauge sheet metal.)

For best results, the sander should be tipped with just enough pressure to bend the flexible disc from the locking washer to the outer edge. If the angle of the sander to the frame is too great, only a small area of the disc will contact the work surface and you'll get a deep, rough cut. If you place the disc flat against the frame, you'll get an irregular and bumpy cutting action and it will be difficult to control the motion.

For frame prepping, the disc sander is used to smooth out torch cuts, welds, and angular corners such as those formed by the addition of boxing plates. It is also used for locating the low spots that result from years of rusting and pitting. When used for grinding, the sander is moved from left to right; when removing grinding marks, it is moved at right angles to the marks.

You also have to know something about sanding discs. There are two types—open coat and closed coat. Open coat discs are designed for paint removal and for grinding down lead and other soft materials, which would adhere to the disc and reduce cutting effectiveness. On open coat discs, the backing is only fifty to seventy percent covered with abrasive grains. This relatively wide separation of grains prevents clogging. Open coat discs are available in grits from 16 to 120. The lower grit numbers are coarser and more widely spaced than the higher numbers.

Closed coat discs, in grains from a very coarse 16 to an extremely fine 320 are designed for work where the ground-off particles offer little or no tendency to clog the disc. The abrasive particles are close together.

In general metal work, a coarse grade open coat disc is used first to remove metal, and then a closed coat, fine grade disc is used to obtain surface smoothness. The accompanying table lists the abrasive grits used in metal preparation according to recommendations made by abrasive manufacturers.

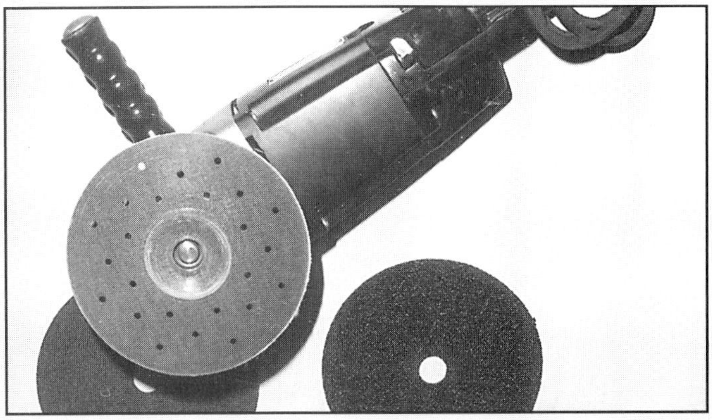

Indispensable to the preparation of a frame for paint is a 7-inch diameter disc sander with an assortment of open and closed coat sanding discs.

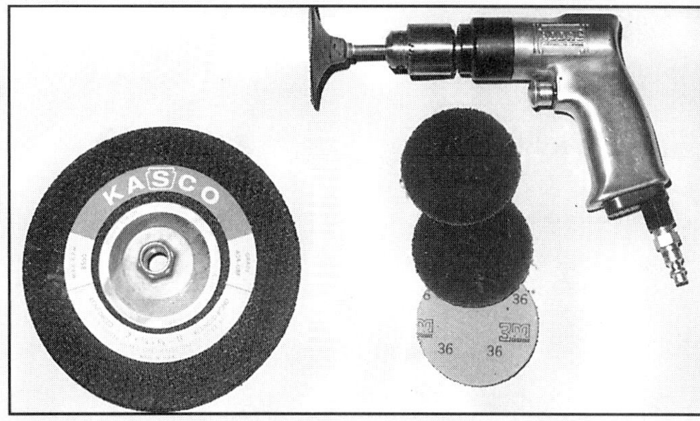

The Kasco resinoid wheel is for the toughest work. Smaller discs are useful where a very light touch is needed.

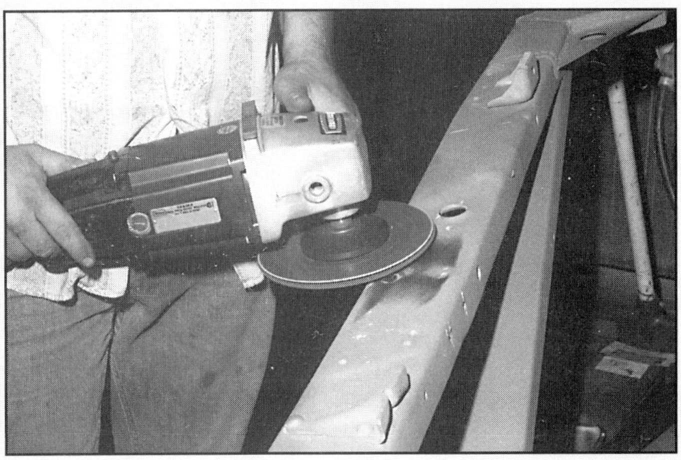

The heavy disc sander should be tipped with just enough pressure to bend the disc, and moved back and forth continuously.

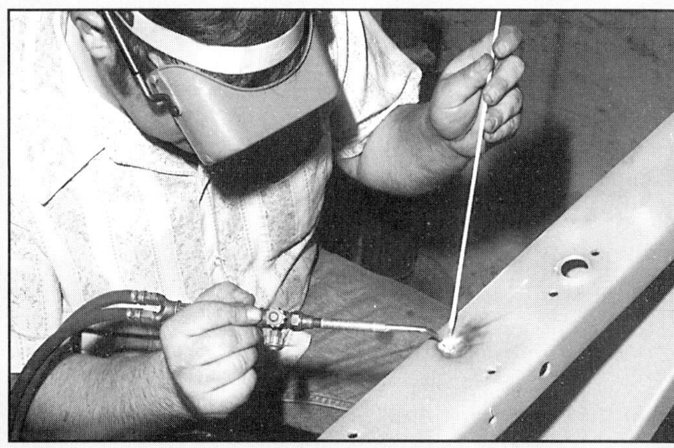

When you lay the disc grinder down, you can search out all the small, unneeded holes in the frame. Brazing is the preferred method for filling them in. And again, instruction pamphlets are available at the welding supply shop. First, however, countersink them to give yourself a beveled area on which to build up material.

STAGE TWO: THE FINISHING TOUCH

There are two products that have proved popular for "final finishing" an original chassis that has undergone modification. One, "all-metal," essentially consists of an aluminum powder in a polyester resin base. According to the manufacturer (U.S. Chemicals & Plastics) and verified by those who have used it, all-metal offers filing and sanding properties similar to those of lead. It adheres to most metals and fiberglass, seals out moisture and sands to a smooth finish.

The second product, Evercoat Polyester Glazing Putty, is the icing on the cake. It is an "ultra-smooth finishing putty that fills minor surface imperfections such as pinholes. It adheres to body filler easily and sands to a smooth finish with no shrinkage." The beauty of the product is that it can be used over paint to feather-in sanding scratches and other minor imperfections.

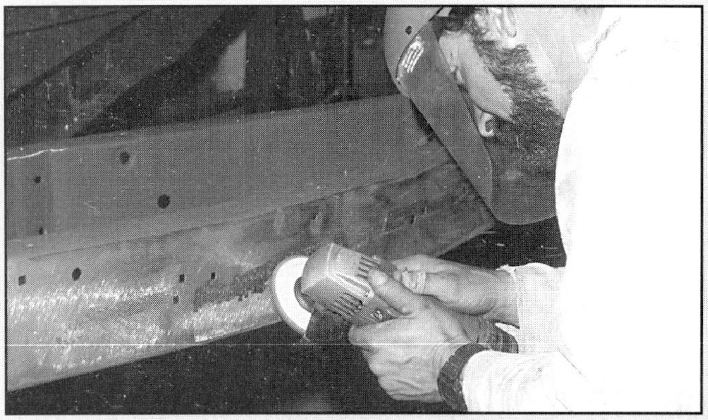
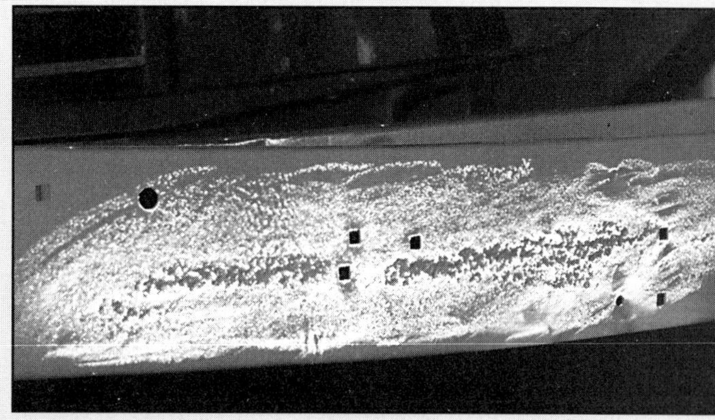

All-metal is used in the early stages of frame restoration, but the chassis rail must be sanded down to bare metal, and kept clean and dry. Depending on previous preparation (whether or not the frame was sandblasted) a closed coat 24- or 36-grit disc is right for this part of the job.

Application is standard for this type of material—place the desired amount of All-Metal on a clean mixing surface, and catalyze it according to the proportions suggested on the container. You don't have to hurry, but All-Metal does set up in less than five minutes in 70- to 80-degree temperatures. It can be sanded and primered in twenty minutes.

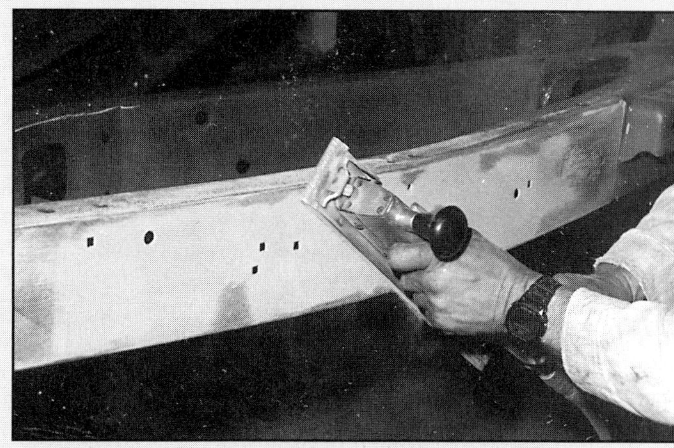

Final finishing (before primering) can be done by hand, or even more efficiently if you have a pneumatic file. Break it with 36- or 40-grit paper, then bring it down to finish with 80-grit, but no finer. Now you use the Evercoat Polyester Glazing Putty to smooth out any minor imperfections.

TECHNIQUES: STAINLESS STEEL TRIM REPAIR

Thanks to an English metallurgist by the name of Brearley, who in 1916 patented a cutlery steel containing 0.70 percent carbon and 16 percent chromium, the automotive world gained a metal that for all practical purposes is forever. It's true; stainless steel alloys resist attack from atmospheric corrosion, organic solutions, hot or cold acids and scaling at elevated temperatures. That's why so many pieces of exterior body trim were stamped from the eternal metal back in the days when automobile manufacturing was a craft and not a cop-out.

Alas, the classics are a thing of the past by definition, and now that costs are prohibitive, painted plastics have taken the place of the beautiful rustless alloys. Be that as it may, even stainless steel will accumulate a lot of dimples and dents during 65 plus years of exposure to road hazards. But hold on, here's some good news. Stainless automotive trim, just as mild steel, can be worked cold . . . even with the simplest of hand tools as Phil King demonstrates. So before you spend big bucks for better swap meet stainless or reproduction trim, try your hand at repairing those items you already have.

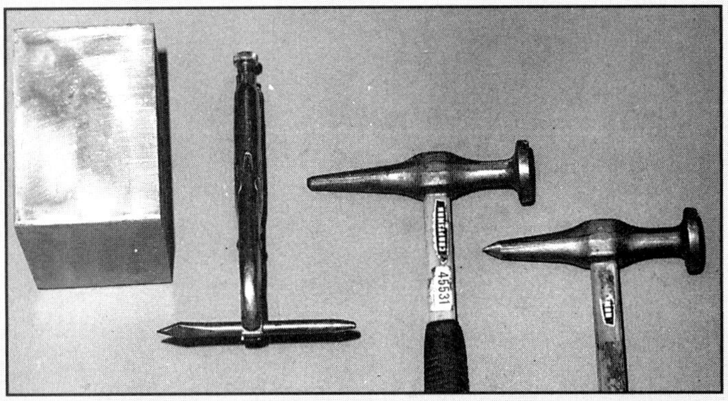

"Picking" and "dinging" hammers are used for bumping out small dents and imperfections. Professional bodywork tools are available, but if you only have a few blemishes to rework, you may just want to make your own out of a broken screwdriver blade and a pair of Vise–Grips. Both a blunt and a sharp point are desirable. Any smooth chunk of metal, aluminum or harder, will serve as a general purpose dolly.

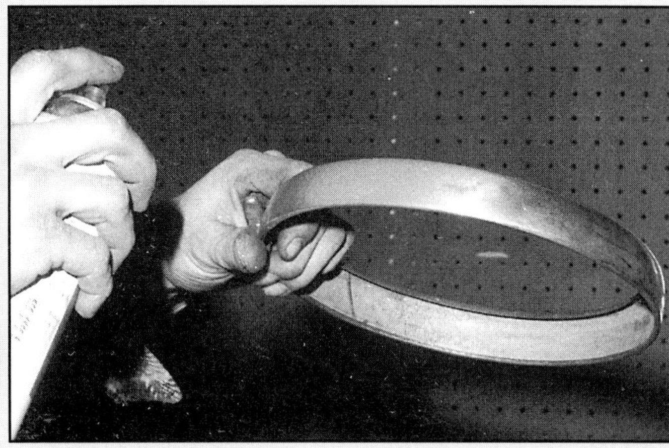

The trick to this kind of metal work is to always know exactly where the high and low spots are. For that, just spray a little primer inside and out for a "guide coat". . .

. . . and lightly sand it off with 320 wet-or-dry paper.

Once the exact parameters of the dents are located, you can proceed to "pick" at them. Work from the inside out.

Use the blunt end of your hammer until you bring the large dents down to smaller ones, then pick in the specific spots. Hit the dent on the high spot with no more force than the weight of the tool affords.

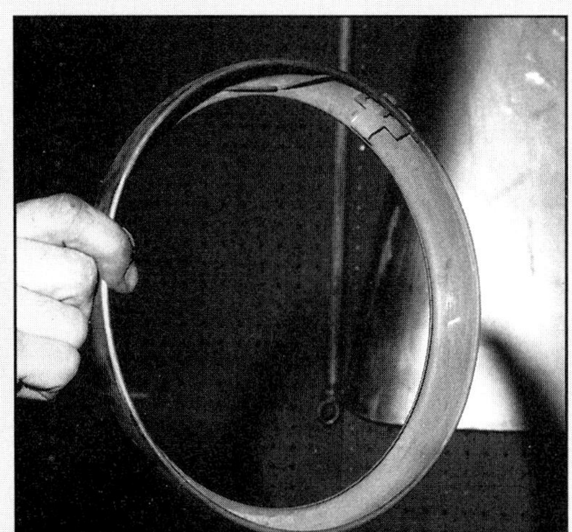

Don't spend too much time on a dent without shooting another guide coat. It makes it easier to see what you are accomplishing. Thin stainless "works" easily, but it is prone to cracking if you pound on it too much. (And, it doesn't take heat well.)

Sears and industrial hardware stores carry inexpensive buffing wheels and polishing compounds. White rouge works nicely on stainless steel and brings it to a bright luster. Do not apply too much rouge to the wheel or too much pressure on the piece. If you do, it will make a gummy mess when the metal gets hot.

With a little patience, (which is easy when you enjoy what you are doing), you can often bring back a piece of stainless trim you previously thought was a lost cause.

TECHNIQUE: HOLE FILLING WITH SOLDER

Since all of you have a soldering iron—and sooner or later, holes to fill, why not do that needed "lead job" yourself and save a few bucks? The basic advantage of soldering over brazing and conventional leading is the panel being repaired is not excessively heated and warped. (The melting point of most tin-lead solders is from 361 degrees to 437 degrees F.)

Soldering itself is quite simple, but let's review the process in general. In order to get solder to adhere to sheet metal, it is necessary to first clean the surface of old paint and rust. Next "tinning" is required to prepare the surface of the metal for adhesion of the solder and to cover the tip of the soldering gun so that oxidation will not occur while it is being used. Acid core solder, melted onto the preheated surface, seems to work best for tinning.

When the tinning is completed, the surface will have a bright shiny appearance and there should be a strong bond between the "tin" and the sheet metal. More solder is then melted onto the area. A flux is necessary to ensure chemical cleaning (the removal of any oil or oxides) and resin core solder is preferred for this phase of the operation. Just enough heat is applied to keep the solder "plastic" while it is spread with the broad tip of the gun. That's really all there is to it, but for a more detailed explanation, take a closer look at the accompanying photos.

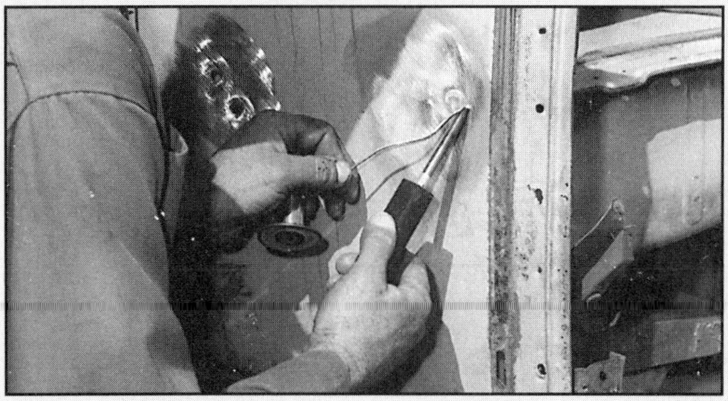

All you need for filling small holes and dents is a standard 250–300 watt soldering gun with a broad tip, metal cutting shears, solder and a coffee can. Holes up to a quarter-inch in diameter can be filled with solder after the edge is slightly beveled with a small rat-tail file for a good bond. Tin the edge, and then work around the circumference of the hole slowly filing in the void.

Lightly grind down the fill-in until it is flush with surrounding metal. If you find any "holidays" fill those in with a little more solder.

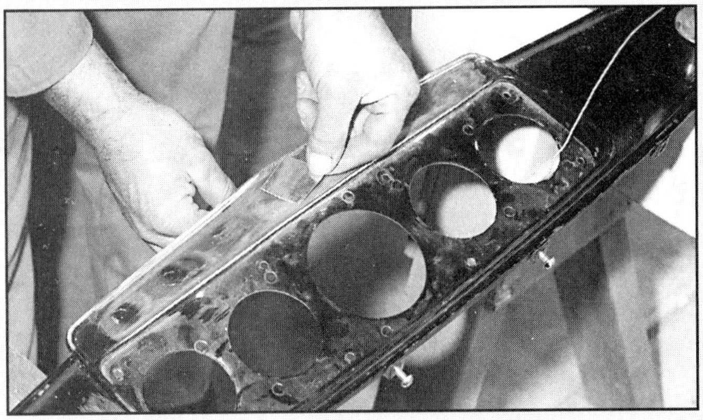

For holes bigger than a quarter-inch, a thin metal patch is necessary. Coffee can metal is ideal, it is lightweight and doesn't draw too much heat. Trace the outline of the hole onto a flat piece. Then taper the edges of the hole slightly so that the patch sits a little lower than the surrounding surface.

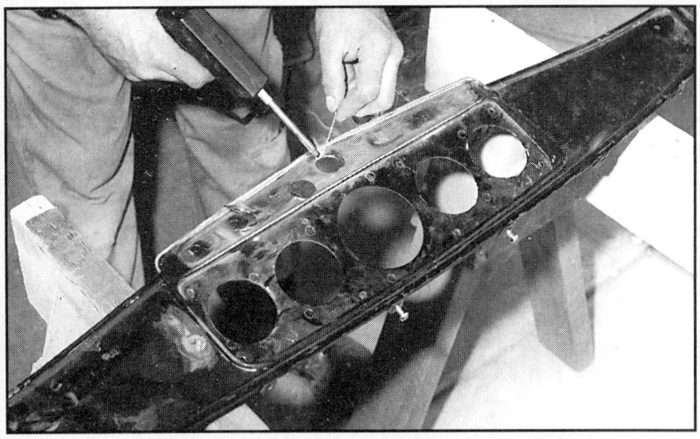

Tin all around the edge of the hole before you fit the plug in.

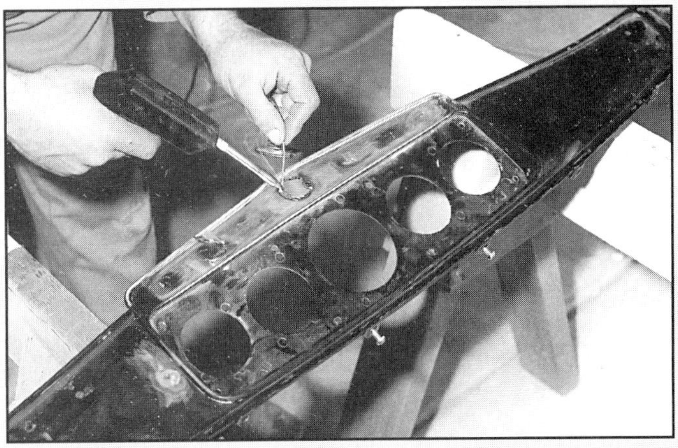

Then simply solder in the patch securely. Resin core solder works best here.

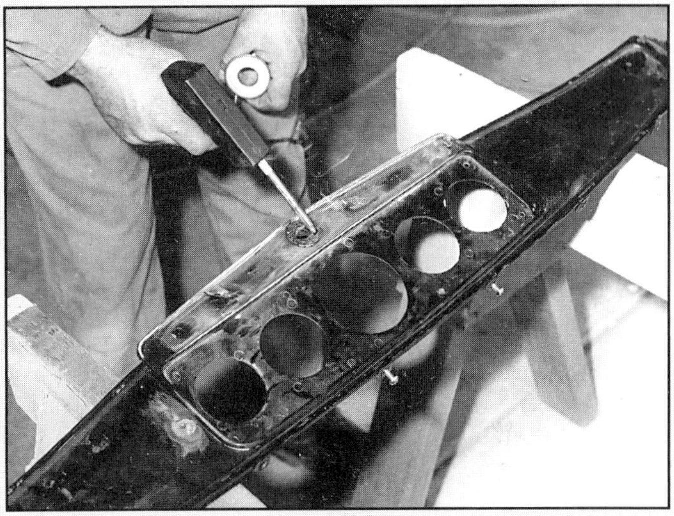

When the joint has cooled, melt more solder over the surface of the patch. Use just enough heat to keep the solder plastic, but not molten. Smooth it out with the tip.

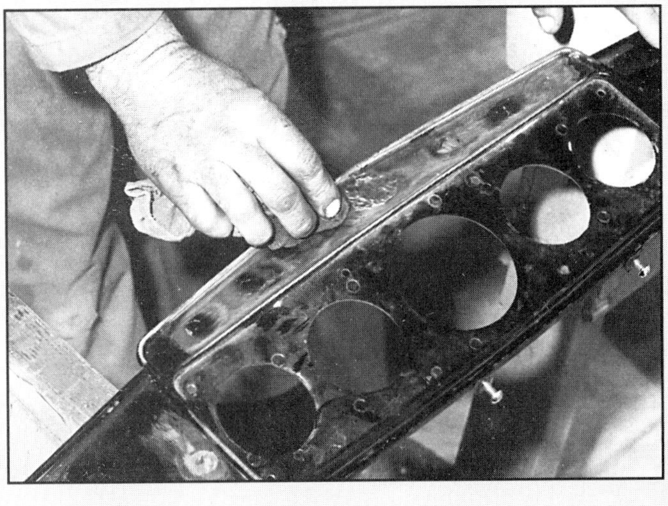

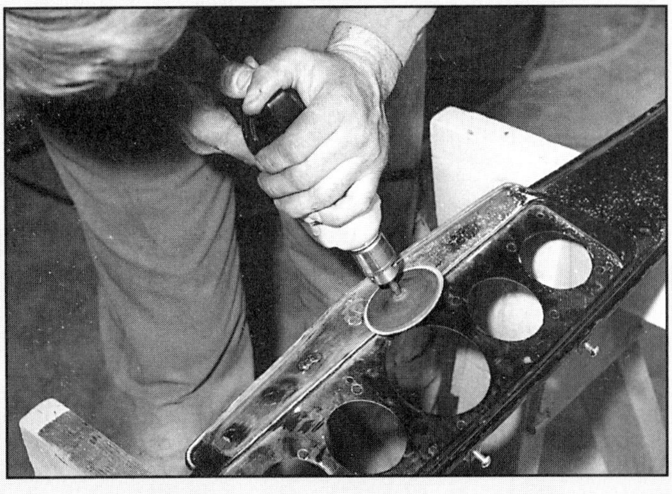

Clean the resin off with lacquer thinner to keep from clogging the sanding disc, then grind down the filled patch with a #36 or finer open coat disc.

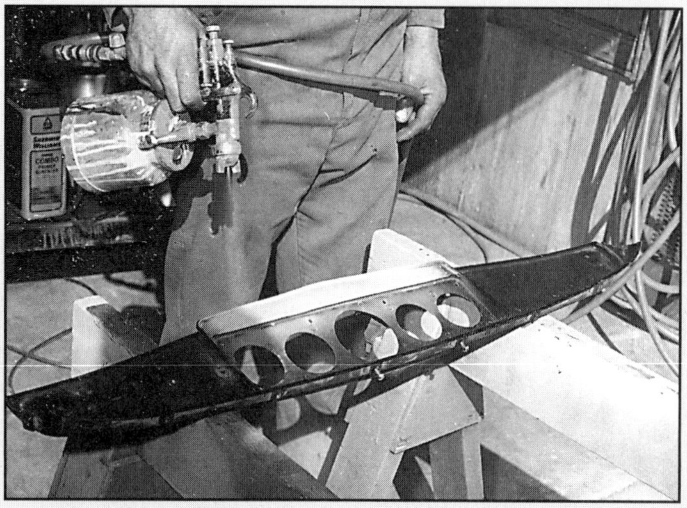

After all holes have been patched and "leaded" in, prime and smooth over minor imperfections with glazing putty. Quick and easy bodywork, eh?

TECHNIQUE: GETTING A NEAR-PERFECT FENDER FIT

Inexperienced street rodders are sometimes perplexed to find that a seemingly simple task such as bolting the rear fenders on an old car can be a problem. Obviously the job begins with top-quality metal work, but all is for naught unless your part (the assembly) is carried through with care.

Some guys like to have their fender welting custom-made by a trim shop in a material that is color-matched to the paint, some like stainless steel welting, and some like the plain ol' black welting off the shelf. That part is a matter of taste. The trick, of course, is obtaining a perfect fit when the welting is wedged into the groove between the body and the fender.

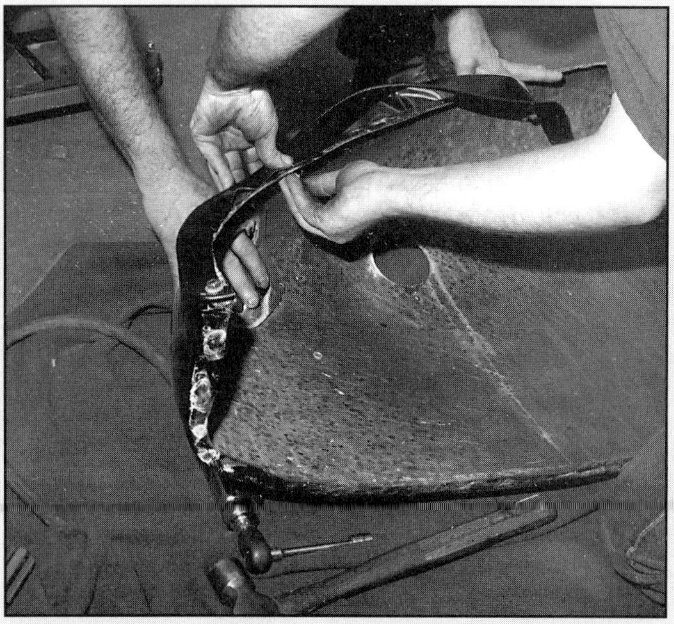

The first step is to prepare the fender-to-body welting. Lay the fender down on a clean moving blanket or similar padded protective surface. Be forewarned it takes two guys to do this quickly and easily; one to hold the roly-poly fender, and one to locate the fender-to-body bolt holes on the welting.

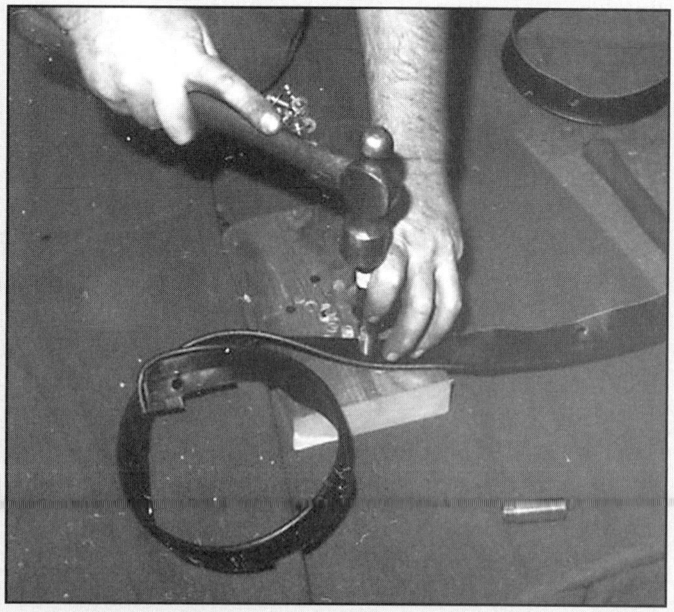

Punch the bolt holes into the welting. Some simply cut an "X" at the mark, but this bunches up the welting. Others try to carve the hole in with an X-Acto knife or a razor blade. This will work, but you had better be right on. A hole punch is best for the job. And yes, the cheaper 3/8-inch diameter punch works, but you'll have better luck with a larger punch.

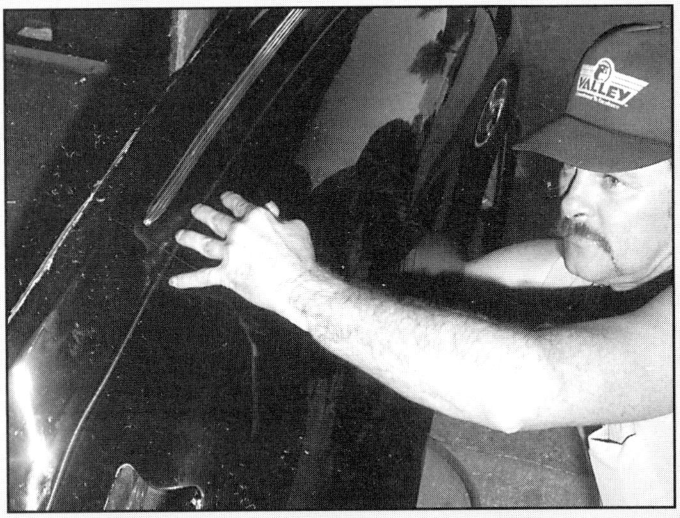

Bolt the fender up to the body loosely with longer-than-needed bolts. Even though this is only a temporary attachment, use rubber-padded fender washers to ensure a squeak-free fit. Properly installing fenders without scratching the paint is just another reason why this is always a two-man job.

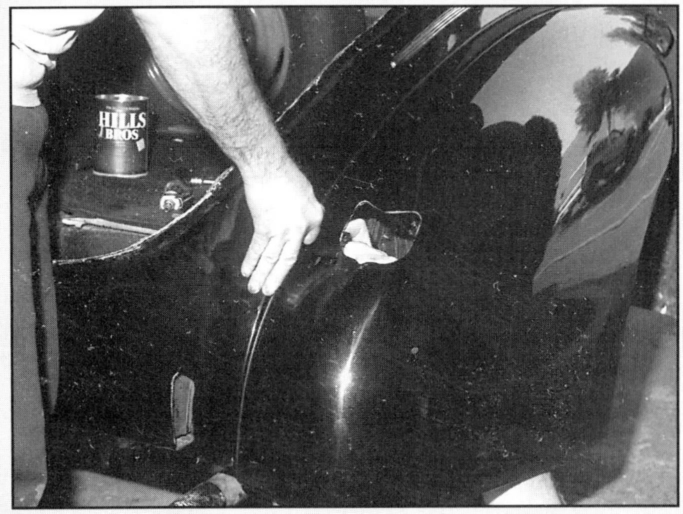

The final fit will take longer than you think, but semi-perfection doesn't come easily. (Few of us will ever live long enough to see true perfection, but we want to get as close as practical.) With a man top and bottom, the fender and welting are coaxed into place. When the "inspector" is satisfied, replace the too-long bolts with those of proper length, and don't forget the lock washers.

TECHNIQUES: A DRIVEWAY SQUIRT

A full-on custom paint job is often an exercise in futility for the average street rodder, but repainting the firewall isn't. It's a well-defined task of reasonable proportions that doesn't require fancy equipment or a spray booth. Not only that, it offers a perfect opportunity to develop knowledge and skill. Before we get to the nitty-gritty of a firewall squirt, however, some precautions and disclaimers are in order. These are the days of OSHA, EPA and VOC. Unless you've been living on Mars, you know something about OSHA and the EPA, the regulatory agencies that oversee the health and welfare of workers and the environment. However, if you're not in the painting trade, you've probably never heard of VOC. Nevertheless, once you put your finger on the nozzle of a rattle can, you are in the mix. "VOC" stands for "Volatile Organic Compound." It covers a wide class of materials used in automotive refinishing products. Basically, it is the amount of evaporative solvent contained in refinishing materials. (Solvent is a major component of not only thinners, but also of the unthinned color and catalyst/hardener used in automotive acrylic urethanes.)

The reason the EPA cares about VOCs is that they react chemically with sunlight. When mixed and sprayed, the VOCs evaporate out of the paint film, and into the atmosphere. There, in combination with auto emissions and dust, they create pollutants known as "photochemical smog." That's why the days of complete driveway paint jobs are long gone.

Well, even though full-bore homebound paint jobs are street rodder tales of the good ol' days, that doesn't mean you can't refinish twenty square feet of classic sheet metal without the comforts of a professional spray booth. It just means you should check with the appropriate local and state authorities in your area regarding air quality regulations. The photo sequence provides the how-to details.

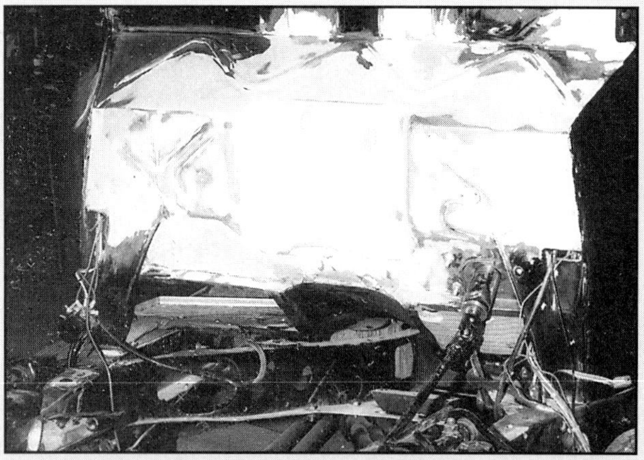

Once a few modifications were completed on my coupe's firewall, the sheet metal was given an expansive, but thin coat of plastic filler. Most of it then was sanded off leaving a silky smooth backdrop for the engine.

Spray guns have changed over the years thanks to the demands of VOC regs. This one is a DeVilbiss JGHV-530 Super Pro-Plus 2. Modern paint guns can still use standard shop air compressors, however. Most are the non-bleeder type, and therefore, trigger the same as older guns. High-pressure shop air is converted to low pressure at the gun itself.

Paint companies have developed a number of low-VOC refinishing products. PPG acrylic urethane primer surfacer is designed to provide fast film build-up with "minimal coats, excellent adhesion, ease of sanding and color-holdout properties." (Color "holdout" means how long the paint color stays shiny!) It can be used over properly prepared existing painted surfaces, and/or primered steel, aluminum or fiberglass.

"Reducer" is just another name for expensive thinner, but make no mistake; you gotta use the good stuff. Mixing paint is like following the recipe for baking a cake; if you don't mix the proper ingredients in the proper proportions, you are going to get something other than what you want. There are several kinds of reducers. What you use depends on the ambient temperature at the time of painting. DT 1865 is used when it is 60–75 degrees F, and DT 1885 is for 75–90 degrees. (We used DT 1865.) It should be noted, however, if it is colder than 60 degrees, forget driveway painting until the weather warms.

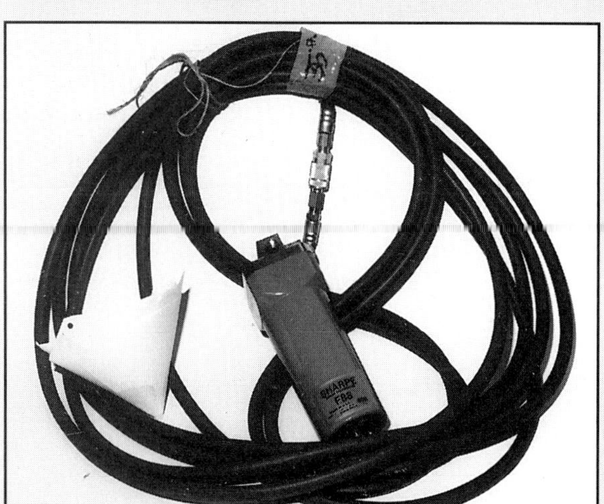

While you're at the paint store, pick up a wad of paper paint filters, several new gallon and quart cans for mixing, and product spec sheets for the specific paint you buy. (There's much valuable information in these sheets.) This is also the time to buy a length of new air hose. Old hose can have air tool or compressor oil in it and you can blow it on whatever you are painting. An air hose oil/water filter is also a must; you want to be absolutely sure there is no moisture in the line.

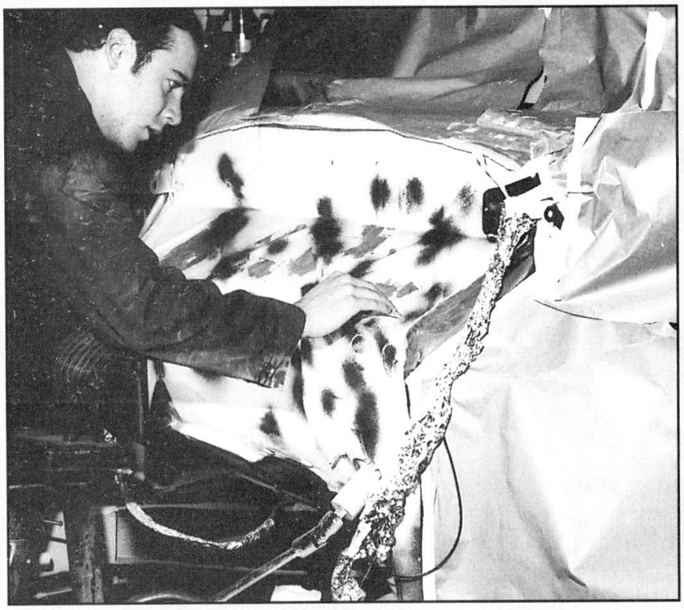

Now, let's get down to business. Tape off the rest of the car to protect it from overspray. It is important to cover whatever you are not painting; paint dust settles everywhere. Tape around edges where the tape line isn't likely to be seen. Aluminum foil is a handy wrap for wiring, hoses and other hard-to-tape areas. Sand the firewall with 220 grit. Although you want tiny scratches in the surface for good paint adhesion, you don't want to overdo it with a grit that's too coarse. Use your air line to blow off the dust, and then wash down the surfaces with soap and water. Re-clean with wax and grease remover. Clean one area at a time, and while the surface is still wet, wipe it thoroughly dry with a clean cloth. This is very important, because if you don't wipe the panel down when it is wet, the grime that the cleaner released will still be there. This is one area where the amateur often gets in trouble—pinholes develop, and the paint does not adhere properly. Use plenty of clean cloths, and discard them after use. Do not touch the panel surface after cleaning.

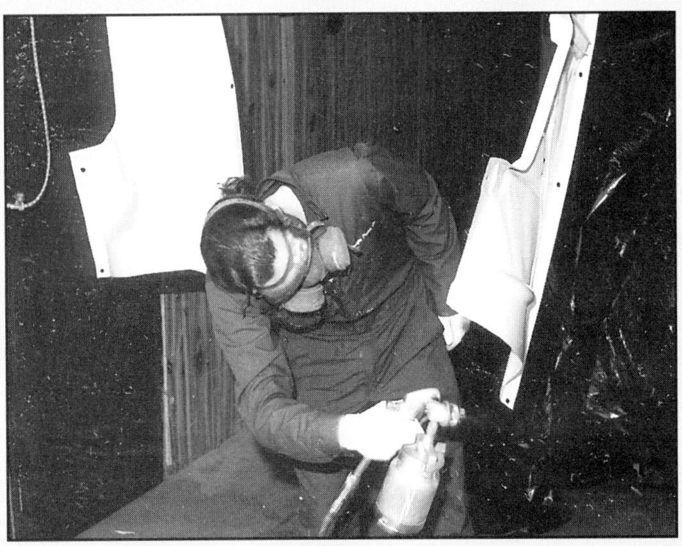

Let the games begin! Angelo decided to primer the inner fender panels next. He mixed one part PPG K 200 Primer Surfacer and one part DT 1865 Reducer. He then added one part K 201. The latter is the catalyst or hardener. If you don't put this in, your paint will be gummy, and it will never dry. The batch was mixed in one of the clean gallon cans, and then strained before shooting. Please note, whatever you do, don't risk your health by painting without a top quality facemask. It is an absolute necessity.

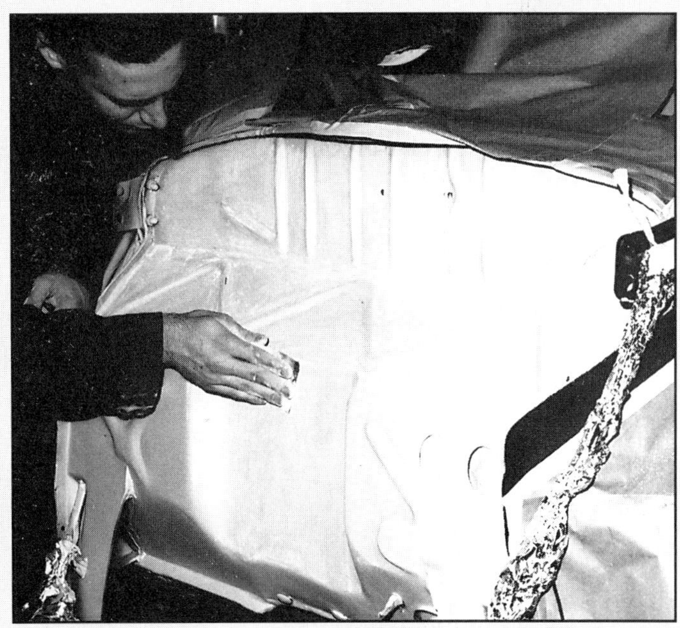

Returning to the primered firewall, he gently sanded it with 320-grit paper. And gently is the key, if there is a breakthrough, you must spot-primer. By the way, do not save catalyzed material. Do one last cleaning with silicone remover before the finish coats.

PPG's Deltron acrylic urethane paint was used for the firewall. It is specifically formulated for use over properly prepared and cured air-dried finishes, and offers fast-drying characteristics for spot repairs. That's something a rodder who is always changing things can really appreciate. Although it can be clear-coated, such was not deemed necessary in this case.

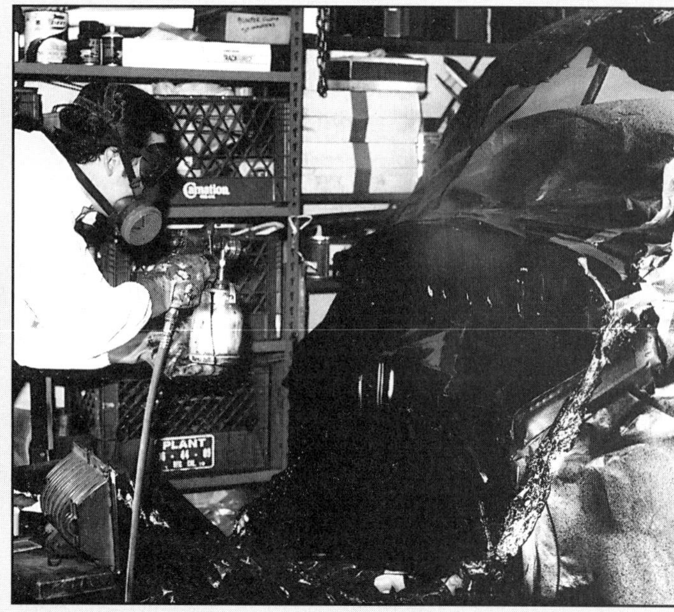

Finally the firewall got some color. Angelo mixed one part black (647) to a half part DT 1865 Reducer. Then he added one part DAU 2 acrylic urethane hardener, and one capful of DX 77 Fisheye Preventer. This minimizes the effects of contaminants such as silicone. The firewall took five light coats, with 15 minutes between coats. The air pressure was set at 40 to 50 psi, and then adjusted slightly higher. Higher pressure equates to better atomization, with less chance of orange peel. If you are not happy with this range, experiment on a piece of sheet metal to find the pressure you prefer.

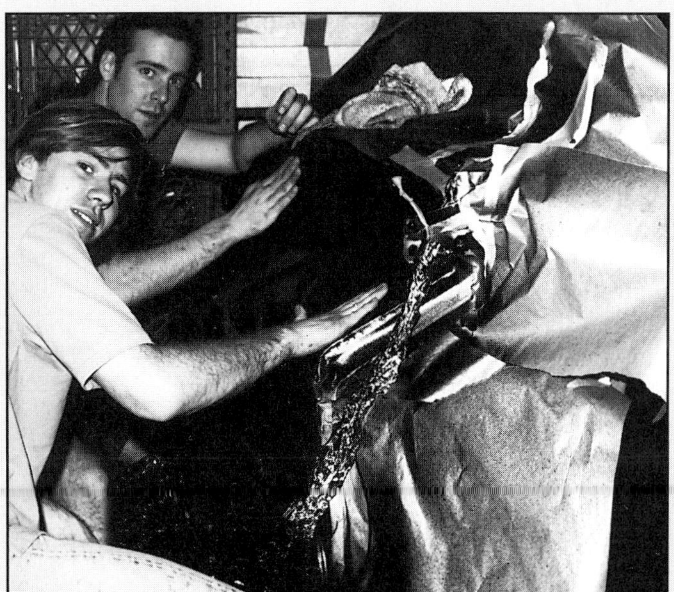

Angelo hit on his pal Dave Chamberlain to help with the 600-grit color sanding, and a bit of touch-up. Count on the latter when a stock configuration firewall is painted. Those reinforcement ribs that look so good are a bear to color-sand without breaking through in a place or two.

The "driveway" squirt turned out well, however, and the engine, fenders and hood were quickly reinstalled. But getting all that sheet metal in sync is easier said than done.

TECHNIQUE: CONQUERING GRILLE GAPOSIS

It's one of those tasks that even experienced owners of 1937–1940 Ford-based street rods are reluctant to tackle. Fewer still are willing to pay a body shop for what appears to be a rather straightforward task: the installation of a grille (reproduction or original) and the realignment of the hood and front fenders into a shining example or pre-World War II automotive Art Deco. The reason, of course, is that it appears to be a straightforward task only to the uninitiated.

Those who have toiled in the street rod vineyards know how difficult it is to properly assemble that mass of sheet metal suspended from the cowl of a passenger car or light pickup. The achievement of a precise, symmetrical fit of the gracefully curved metal sculpture is a challenge to say the least. Even savvy rod builders have been known to hesitate when faced with the problem. And only then have they been forced to admit that maybe the body shop estimate of ten or more hours of labor was not so far out of line. To paraphrase one of my favorite racing axioms, perfection costs money, how much perfection do you want?

When you can no longer ignore the sad-sack look of your rod's front-end appearance, call on an eagle-eyed rod buddy to donate a Saturday (and maybe part of a Sunday) for a needed helping hand. Be prepared, however, to find several well hidden "dirty little secrets" when the prow of your pride and joy is torn asunder. Sixty-five years of road service, fender-bending and hood clash, coupled with repairs made by less than persnickety body shops are very likely to make a surprise (and unwelcome) appearance.

I was satisfied with the fit of the hood and fenders on my coupe, but the "grille gaposis" was starting to annoy me. (And yes, if you look closely, you'll see that I once again busted the lower grille "chin" in a moment of carelessness.)

I long had eyes for one of Vintique's reproduction grilles and finally, I decided to go for it. It's a work of reproduction art. I also picked up a new grille chin, a center plate, and an installation hardware kit while I was at C. W. Moss in Orange, California.

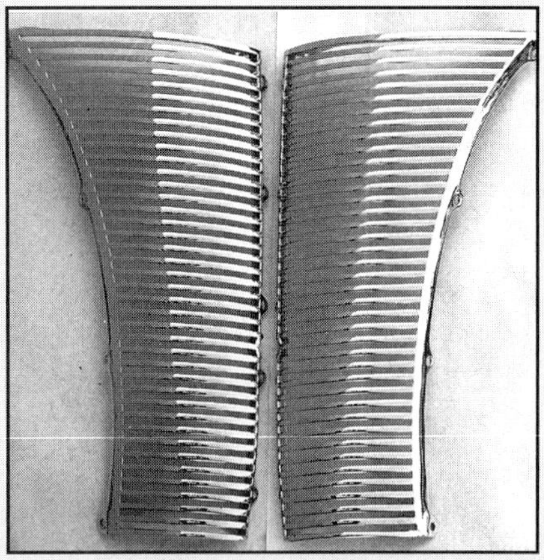

It is well understood by 1937–'40 Ford rodders that only after a near-perfect installation of the hood can you begin the replacement of any remaining front-end components. The hood must be bolted on and as married to the cowl as possible. Make your inspections and necessary repairs before you begin the assembly.

John Barrera bolted the die cast Vintique grille, newly painted chin, and newly chromed center plate loosely together on a blanket-covered piece of foam rubber in the conventional manner, using a drift punch to align the holes. The proper nuts and bolts came with the Vintique grille hardware kit.

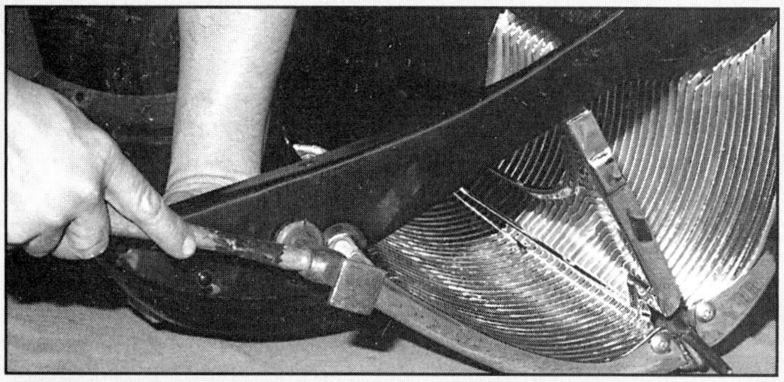

The coupe's louvered sheet metal grille sides are original (although excellent reproductions are currently available). The originals, however, had to be "hammer-fitted" to the die cast sections. (Don't get heavy-handed here!) At this point John noticed two more-or-less fresh cracks in the original metal. It brought home one of the truisms of this piece: a street rod fashioned from a decades old steel body is going to require ongoing inspection and readjustment as well as repair if it is to be kept in a solid, squeak-free state. Metal fatigue occurs in dozens of stress points.

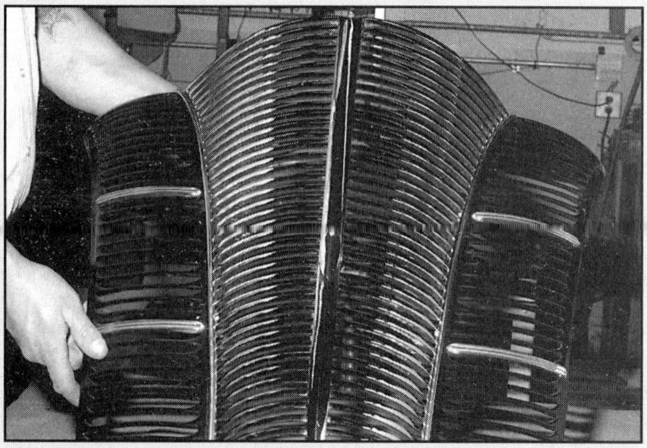

Once welded and repainted, John reconstructed the grille assembly. He followed an empirically determined assembly sequence, going from left to right, top to bottom, finger tight to snug.

The fender welting and outer grille plates that deflect air to the radiator are not easy to tuck in when, at the same time, you are trying squeeze the grille assembly between the fenders. One solution is to pop-rivet the plates and welting to the fender interface.

The grille was then shoehorned in and the entire front end bolted together, snug but not tight. The front fenders were next loosened from the cowl. A padded floor jack was rolled under the grille, and used to gently work the entire assemblage up to meet the hood.

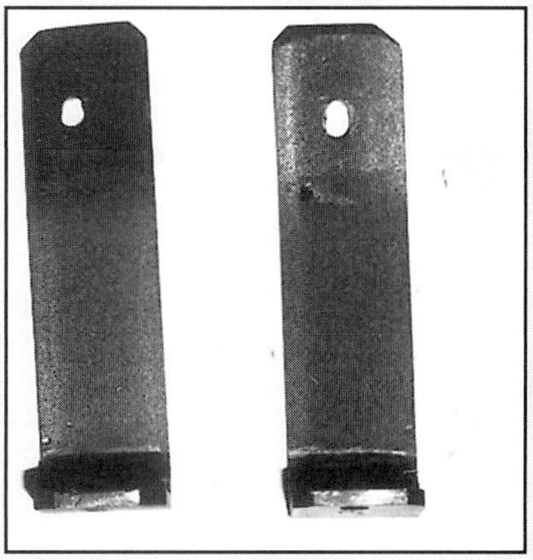

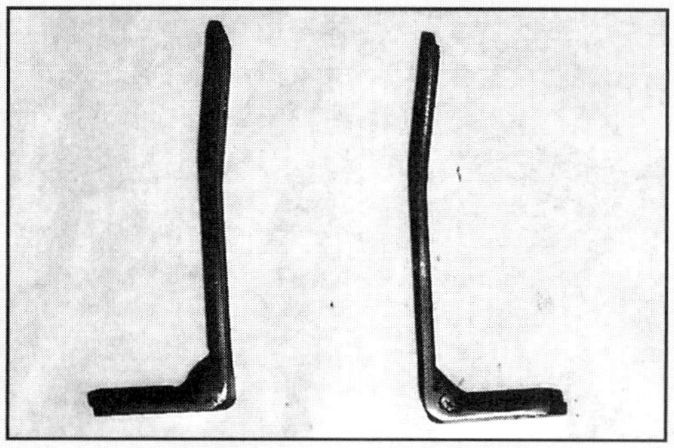

Various types of home-made fender braces have been used through the years by '40 owners to replace the oft-removed original one-piece fender brace mounted to the front crossmember. In place are two hefty quarter-inch steel brackets with gussets to immobilize the entire front assembly. They are bowed slightly to bring the fenders in for a better fit to the coupe's modified chassis. The ones you construct for your car would not necessarily have the same contour, but these are 7 1/2-inches tall and 2 inches wide with a 2 1/4-inch base. The hole for the fasteners was located while the front assemblage was jacked into place. Again, and this is critical, the hood ornament is the alignment reference for both fenders and grille. With the grille and sheet metal cinched down, the gaposis of the old grille was a thing of the past.

Unfortunately, I just wasn't satisfied with the fit of the back of the hood at the cowl!

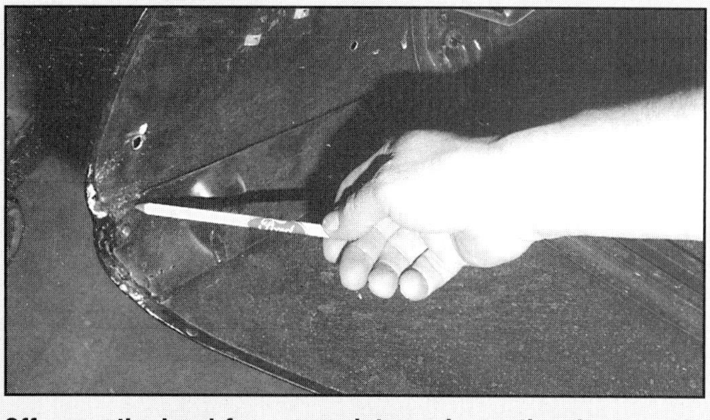

Off came the hood for a more intense inspection. It was only then that we discovered the car's dirty little secret: the hood's inner bracketry had been modified sometime in her past! I owned the car for more than 30 years, but in all that time, I had not paid attention to the modification!

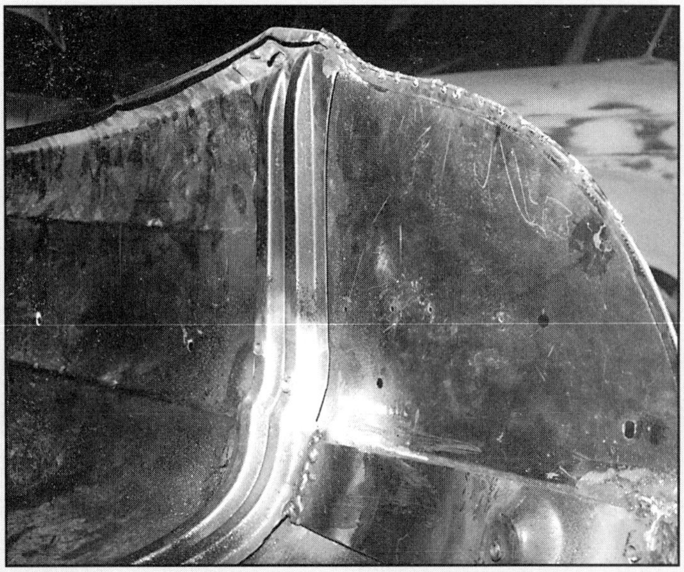

Suffice it to say that the old hood fix was causing this new misalignment. The previously repaired bracketry was removed, repositioned and the pieces welded back in, and then a little metal was added to the back of the driver's side of the hood to achieve the hood-to-cowl fit that I wanted. I must admit this repair is too complicated for an amateur to tackle. I do not think the problem was simply metal fatigue, but rather some damage in the old coupe's checkered past prior to my acquisition. Inasmuch as the hood is the only "moving part" in the front end, it should be inspected before realignment begins.

As the metal repair of the hood progressed, the ease of opening and closing the latch mechanism stiffened. The problem was traced to warpage in the original hood crossmember (Ford PN 01A-16774-B). Two flat washer spacers between it and the inner lip of the hood (hidden discreetly within the assembly) repositioned the crossmember a tad higher. The release operates smoothly, with just the right amount of effort required to securely clamp the hood down. If you have the same problem, give this quick fix a try.

For years I've heard guys complain about the '40s bosom. Sometimes the problem can be adjusted out, sometimes not. The foregoing amply illustrates the fact that street rodders cannot afford to forget: decades-old cars acquired with the body and frame in a stock or near-stock state almost always bring along some hidden baggage. It's easy to identify what former customizers or hot rod builders did to the body or chassis. It's usually more difficult to ascertain what a mid-century collision shop did. Repairs that were adequate fifty years ago may not be tolerable today!

TECHNIQUE: REBUILDING EARLY FORD DOORS

Next to kicking tires, one of the best ways of gauging an old car's condition is by closing its doors. That's why it is always a treat for a street rodder to hear his passenger remark upon exiting the car, "Man, these doors are right on!" But then, maybe your door-slamming friends have been bad mouthing your 1928–'48 Ford's hatches, telling you to go down to the automotive dismantler and buy some late model electronic entry and exit hardware . . . and, oh yes, pay a pro to graft them in. Easy for them to say, regardless of whether or not the latest jelly bean's pulls would be a cosmetic abomination on your resto-rod. Here's an alternative with a painless price tag that you can do yourself on a Saturday.

Assuming your early Ford has at least a decent anatomy, occasionally the door will jump into the lock mode when it is slammed too hard. Most often this is due to a worn detent plate or a weak spring. The spring must be "live" enough to hold the remote control link and detent plate steady. The second door hang-up is found in the locks themselves. There are two small flat springs that control the overall smooth operation of the lock assembly and as they weaken with age (or break) the handle action gets sloppy and door is hard to shut. Replacement of these springs is a relatively simple task, once you know how. And, fortunately, most of the required parts are available from the vintage auto parts emporiums scattered all around the country.

A worn detent on the remote control link assembly (Arrow 1) is a bit of a problem to repair. Not available new, they must be salvaged from a parts car. A weak or broken spring, however, (Arrow 2) is a simple replacement. The retaining housing holds the spring and also doubles as a bearing surface for the door-handle shaft. To remove it, gently bend back the tabs, taking care not to fatigue them.

To remove the old cup assembly, rotate the large regulator gear to a position where the stock rivets can be drilled or ground out.

Although the repair kit comes with steel rivets, it is much simpler to pop rivet in the replacement. Apply grease to the gears and moving parts. Test for free movement of the regulator before reinstalling it in the door. The replacement springs for the door lock assembly are in constant use, and eventually they break or weaken. But, by and large, that is all that goes wrong with the early Ford door lock. Replacements are available from C. W. Moss in Orange, California and many other antique auto parts emporiums.

The lower spring is held in place by the small "U" shaped retainer. Gently pry back the tabs and remove it.

Replacement of this spring is quick and easy . . . if you don't break the retainer.

The replacement of the upper spring is slightly more difficult. It is held in by a shouldered rivet (left of the assembly) which is not available. But it can be reused if you carefully turn back the edges to get it out intact. Next, pry up the slide and remove the worn spring. With the spring replaced, insert the shouldered rivet and crimp down the edges.

Want to go further in rebuilding your doors? C. W. Moss, et al, has new door checks for all early Fords and replacement key and lock assemblies for 1932–'48 Ford and Mercury and 1932–'52 Ford trucks.

TECHNIQUE: DECK LID SEAL INSTALLATION

One of the most trying installations is the long rubber deck lid seal used on the cavernous coupe trunks of most marquees from the mid 1930s through the late 1940s. Although the rubber is usually installed in two or more sections, keeping the long strip(s) in place during the installation and cement drying time is difficult at best, and costly in terms of wasted product at worst. (To say nothing about leaks!) Try this method.

Drop by your favorite auto parts store and look over the selection of rubber-to-metal adhesives and sealers. There seems to be one for every possible application, but I've had good luck with the plain old black weatherstrip stuff. And yes, get a quart of adhesive cleaner. You're gonna need it!

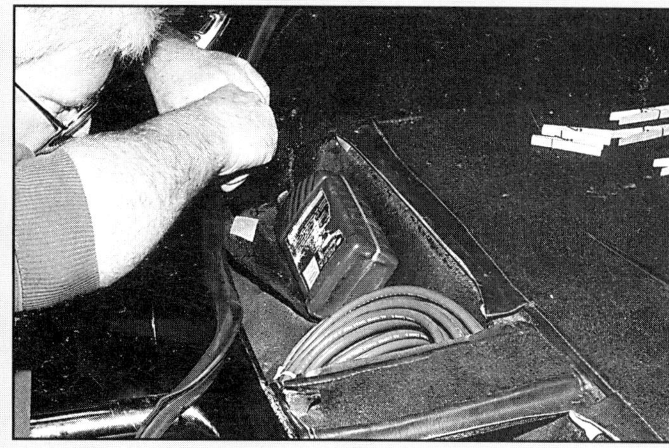

Apply the adhesive as per instructions, and avoid the tendency to put on too much.

Secure the rubber to the sheet metal lip with spring loaded clothespins.

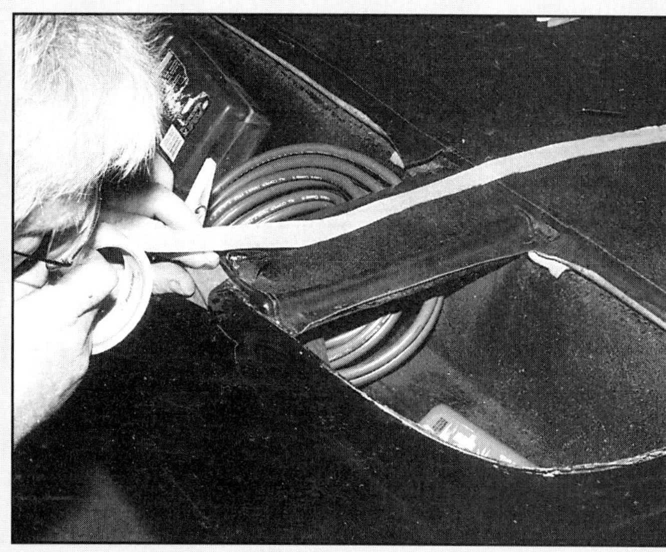

Use masking tape to hold the ends of the rubber in place, and whenever possible, locate the rubber joints at the lowest part of the trunk opening. When the cement has "tacked up," remove the clothespins, carefully close the deck lid and immediately check the fit while there is still time to make minor adjustments.

TECHNIQUE: BUMPER AND ACCESSORIES INSTALLATION

Assembling old car bumpers, brackets, grille and bumper guards can be a bear for a fellow working alone. More often than not, installing them as a tightly assembled unit leads to frustration and assembling them loosely tightened adds insult to injury—scratches on a freshly painted fender. Here's the most successful method I've found . . .

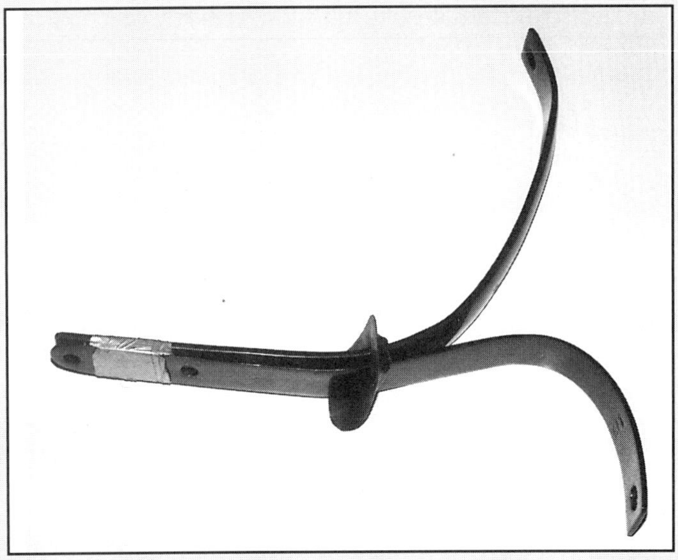

First install the rubber grommet on the paired bumper brackets. A little WD-40 makes it slide easily. Next, align the rearward bolt holes on the pair and tightly bind the brackets together with duct tape.

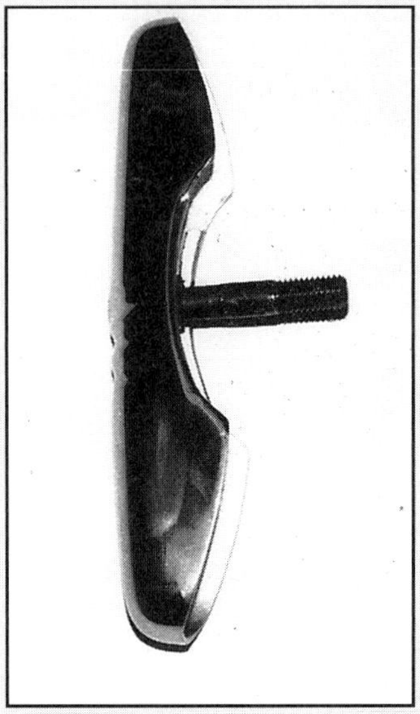

Now chase the threads on the captured nut in each bumper guard and squirt in a few drops of machine oil. Run a stud into the bumper guard. (It will remain in the guard and serve as a fastener instead of a bolt.)

Install the bumper brackets loosely in the frame, and then hang the bumper on the outboard ends of the left and right bracket pairs. Run the nuts on each end finger-tight. Then, with a big, padded C-clamp, draw bracket and bumper together while inserting the studded bumper guard and grille guard. Leave everything finger-tight until you're ready to cinch the entire assembly to the frame.

Chapter 10

INTERIOR TIPS

Got leather? Special cleaning soaps often dry out the classic upholstery material, rather than keeping it supple. It's also difficult to avoid rubbing surface grime into the grain. Neat's foot oil, available in leather goods stores, is the principal active ingredient in most leather-treatment liquids. It is preferable to hand rub neat's foot oil into the leather, allowing twelve hours for full absorption. Then all you need to do is buff it lightly with a clean dry cloth. For routine cleaning, ordinary liquid detergent (diluted), applied with a medium soft scrub brush is preferred. A wipe with a clean damp cloth ends the scenario.

Old-time hot rodders liked Stewart-Warner boat instrument panels for their dashboards. They were offered in three or four different faces but the engine–turned style dominated. Today's traditionalists still love 'em, although many rodders want a dash with more individuality. To meet the need, several companies offer custom instrument panels.

MAKE A PATTERN

To get an idea of what is practical for your car, arrange the gauges of your choice on the bench the way you like them. Then, using a drawing compass lay the arrangement on a sheet of poster board in full scale. Trim the poster board to a manageable size and attach to your existing dash with masking tape. Sit behind the wheel to get a rough idea of how your proposed pattern will look.

ROOM WITH A VIEW

Can you easily see the primary gauges (speedometer, oil pressure, cooling, fuel level and battery condition) without

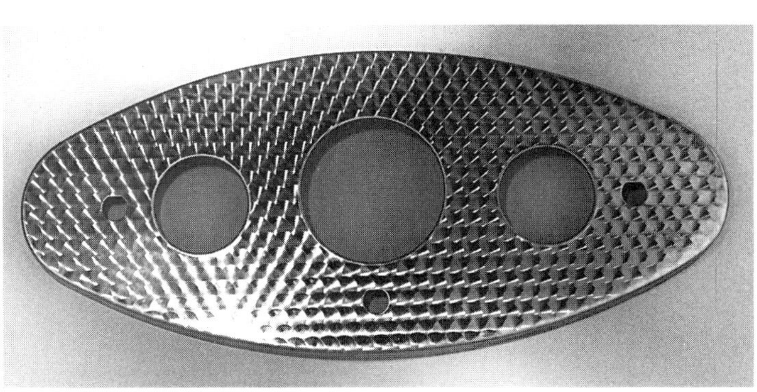

steering wheel interference? Can you see the secondary gauges (tachometer, vacuum, and so forth) comfortably? If not, juggle them around until you can.

Gauge placement is important in homemade cars; you must be able to monitor them all with just a glance. Since street rod dashboards are on the smallish side, if you can squeeze in the primary gauges, don't sweat it, you're off and running.

When you're satisfied, make up a fresh poster board mockup. Don't cut the holes out though; just draw the gauges and note the center points as well as the location of ignition and light switches or whatever else is needed on the dash. Do it neatly and send it to the manufacturer for a quote.

"A" PROBLEMS

Fitting aftermarket gauges in the original instrument panel of the Ford Model A cowl/dash is always a challenge because it was part of the gas tank. The trick is to use a flat 1932-style dash made to bolt in the Model A. Several brands are available in both fiberglass and steel. One design is for closed cars and another for roadsters. (By the way, it is important to remember the builder of a Model A roadster needs the dash and the header plate.)

A MODERN TOUCH

Dobie Gillis, the nom de plume of a *Street Rodder* magazine correspondent, once advised us that the late 1970s Mustang II instruments were six-volt with the same basic mounting as gauges from Ford's post–WWII models. It will take a little trimming and shimming, but the symbols add a subtle touch to a resto-rod. The Ford manual calls for an 8.5-ohm resister in the circuit for each gauge.

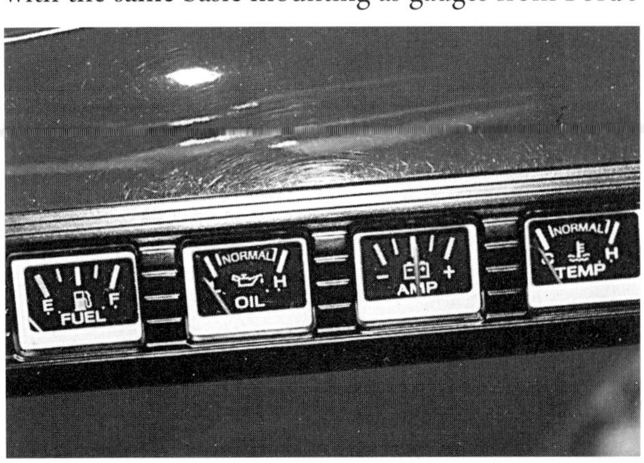

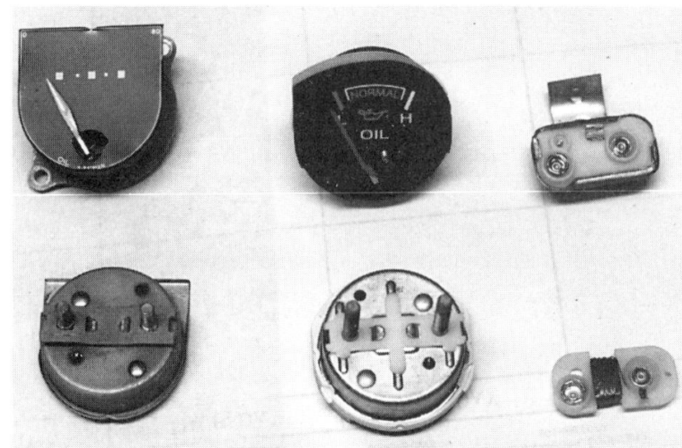

MODERN COMFORT

Most Model As are short on ergonomics, however, late 1980s/early 1990s Chevy S10 Blazer seats tilt and slide forward once they are released, and are a decent fit in Model A Tudor Sedans.

ONE BUCK BRACKET

Mounting a new steering column to the dash in an unobtrusive way can be doubly vexing when one doesn't want to disturb the existing sheet metal and wiring. Experiment with an exhaust tubing U-bolt. There are a number of diameters available, and when painted in a matching color, are about as "invisible" as one could want.

LOUD SPEAKERS

Speakers mounted in doors or cowl kick panels eventually vibrate loose. Often installation screws do not anchor the speakers securely to the conventional fiberboard panel. However, small plastic anchors for use in gypsum wallboard will often work.

VEXING VIBES

Automotive interiors contain large areas of flat metal, to whit, the floorboard, firewall and fender wells. Sheet metal reflects and transmits sound efficiently and can even contribute its own noises by "oil canning" or vibrating back and forth. Most interior noise comes from air turbulence and unavoidable engine vibrations transmitted through the chassis and body.

Since unpleasant sound can drift through tiny openings, the first step in reducing interior noise is to make the interior as airtight as possible without eliminating ventilation. Check all door and window rubber moldings and seals, and replace where needed.

Pull up the carpets and inspect for any holes in the floorboard or panels that may have been overlooked. Seal them off with silicone rubber. Also pack a little silicone rubber around any wires or plumbing that pass through the firewall.

"Q" FOR QUIET

High-frequency interior noise is reduced by plush cloth upholstery, thick headliners and soft fibrous soundproofing such as jute padding. There are other sound-deadening materials on the market as well. One of the best is a product called "Q-Pads." These are adhesive-backed, waterproof, 16-inch square panels originally designed to reduce noise and vibration levels in building floor applications. Q-Pad No. 117 is 0.165-inch thick and can be applied in layers to build additional thickness. It is cut to the desired shape with shears and can be heat-softened (up to 300 degrees) to mold to irregular surfaces. Q-Pad No. 116 is for application on vertical or inverted surfaces, i.e., doors or the roof. Both varieties are available at auto paint supply stores.

KEEP YOUR COOL

One of, if not the largest automotive heat shield manufacturers, is Thermo-Tec Automotive Products (P. O. Box 96, Greenwich, Ohio 44837, 800-274-8437). Their website (http://www.thermotec.com/products/full/full.html) lists no less than 28 separate products from Exhaust Insulating Wrap to Super Sonic Acoustical Mat.

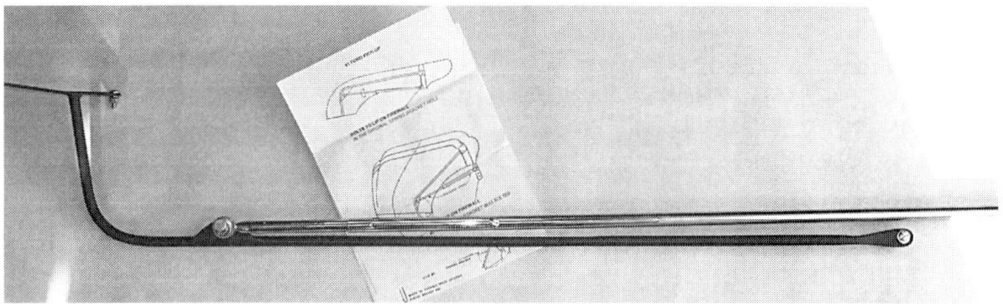

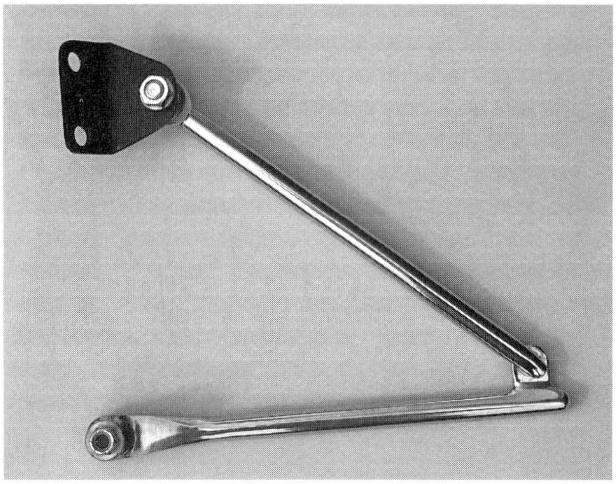

A HELPING HAND

With 1937–'38 Chevy coupe deck lids as heavy as they are, builders may want to install gas struts to assist in the lift. If so, get Maremont Strong Arms (PN 4800). They are a direct replacement for the stock support arms. Just drill out the rivets securing the originals and the struts will fit in the factory holes. The length is exactly the same, and after you lift it halfway, the deck lid will go up the rest of the way by itself. Metric nuts must be used on the strut studs.

For those of us who own 1939–'40 Ford coupes, and prefer "never-fail" mechanical support bracketry over spring-loaded, hydraulic, or gas-filled deck lid lifts, etc., check out this simple, but attractive support bracket. Dave Reggio of Hyde Products (1041 Rue Verand, Slidel, Louisiana 70458; e-mail hydeprod@bellsouth.net) builds them as well as stainless steel mechanical hood holders for a variety of 1939–'41 Ford years. I have had very satisfactory experience with one on my coupe and another on my '41 pickup.

Chapter 11

COSMETIC TIPS

Street rodding is not just a hobby of the mechanically minded, although most of us are "gearheads" through and through. There's an easily identified touch of exhibitionism in our pursuits. In fact, a greater percentage than you may suspect start every Saturday with a polish cloth and a can of wax. More power to 'em, and here are few tips that some may have overlooked.

'POXY PAINT

The most durable protective finish for a chassis and many steel components is electrostatically applied thermoplastic epoxies . . . or powder coatings for short. Epoxy particles are fed from a tank to a handheld nozzle, where a low amperage/high voltage electrostatic charge is applied in an ionized atmosphere. The charged powder is sprayed on the electrically grounded part, and then baked for 20 minutes at 375 degrees. The coat is of uniform thickness and tough enough to withstand a 160 inch-pound impact. Check with your local powder coater for the latest in colors. And, if you also want to keep those new aluminum wheels looking good ask him about baking on a clear finish. Powder coatings resist many mild acids and caustics, but not plain ol' brake fluid. Clear powder coating is also a great way to keep that brand-new intake manifold looking as good as it does when

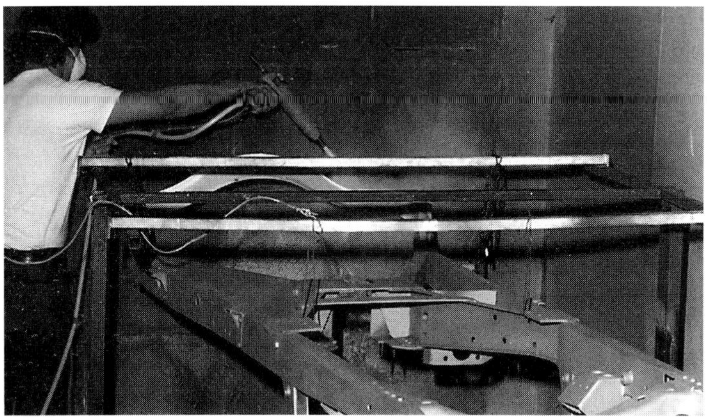

it comes out of the box. The resulting smooth finish is easier to clean than the as-cast surface.

Another great way to protect (or dress-up) aluminum wheels and/or other aluminum components, is via anodizing, an electrochemical process that controls surface oxidation and corrosion. Your choices range from clear to a rainbow of colorful dyes.

CORROSION RESISTANCE

Have you folks down in the rainy Southeast or up in the snowy North with salted roads ever installed black Allen-head machine screws only to find that the wrench pocket rusted away? One obvious solution is to replace them with stainless hardware, but when a black finish is desired, coat the heads and pocket with clear fingernail polish after they have been replaced.

ENGINE PAINT

Rodders often devote much thought to deciding what color they should paint their engine. Hint. If you have a green car, don't paint it red; it'll look like a Christmas decoration! But that's not the real problem. Durability is.

The key to a long-lasting paint job on an engine is in the prep. If you are having the engine rebuilt, the heads and block are always hot tanked before the machine work, and often painted after the machine work. Ask the shop to leave them unpainted, as you'll do that part yourself. Before the heads and pan are assembled on the block, scrub them down with hot soapy water and a wire brush. Repeat the process with a wax-and-grease remover. When the parts are dry, carefully mask off the machined surfaces. Trim off any selvage with a razor blade. It's easiest to spray the block on your engine stand.

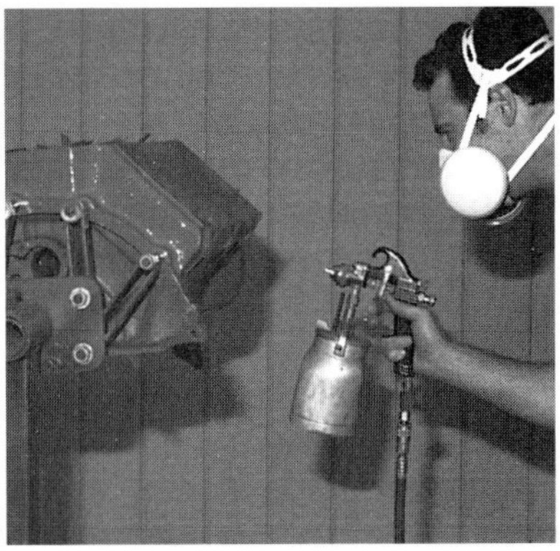

ART WORK

The cylinder heads are best painted by brush. Mask the innards off as above. A great engine paint is "One Shot Enamel" used by pin stripers and sign painters. It doesn't flake, crack or discolor for years even on hot cylinder heads. Brush on two unthinned coats 24 hours apart.

COLOR CHOICES

As for the color . . . well, we all have our preferences, but first review the section on cooling. Of course, most any color goes with a black exterior. Next to that is the same color as the exterior. Also, variations of the body color often works well. That is, a dark green engine in a light green car is "clash-free." Beyond that, you're on your own. And who said Christmas colors are unattractive? Regardless of the color of the engine, replacement of plain-Jane nuts and bolts with polished stainless steel is a nice touch, but again, only if not overdone. Always avoid the "Frankenstein Look."

USE YOUR IMAGINATION

Although there are a number of universal aftermarket engine dress-up items available, dedicated kits aren't available for all the engine makes and models street rodders may install. So, as in the early days of street rodding, some of you will have to design your own "kit." Stock air cleaners, rocker arm covers, pulleys, factory brackets and accessories can be polished, chrome plated, cadmium plated, copper plated (and then top-coated with polyurethane), aluminized, powder painted, anodized and even porcelainized! Let your creative imagination tempered by tastefulness tell you when to stop.

MR. CLEAN

A good way to clean aluminum components that haven't been polished is with one of the alloy mag wheel cleaners, but don't get it on wiring or gaskets. Some cleaners contain diluted acid. Before you get carried away with enthusiasm, find an unobtrusive spot to test various brands on polished aluminum parts. To remove those especially stubborn gas stains on an unpolished, veteran intake manifold, use a stainless steel welder's slag cleanup brush with the alloy wheel cleaner. (Always wear eye protection.)

If such attempts fail to please you, it just may be time to take that treasured manifold to a shop that does abrasive bead blasting. Just be certain they mask off all the machined surfaces.

CAR-NEW-BA WAX

Early pitting on re-chromed bumpers and other non-stainless steel exterior components can frequently be traced to shoddy workmanship. On the other hand, if your items were plated more than three years ago, the pitting is likely due to normal oxidation. (Few chrome shops are willing to give you a written guarantee of lifespan for a number of valid reasons.) So it behooves you to keep the rust gremlins at bay as long as possible. The quick and easy approach is to apply plain carnauba wax (not cleaner wax) after each wash.

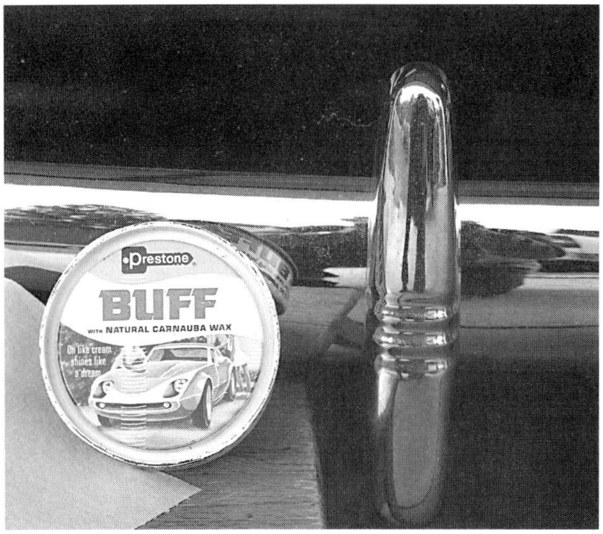

WAX ALTERNATIVES

Another protective measure is one of the aerosol cans of clear polyurethane. There are several name brands, and some are more durable and less prone to yellowing than others. My best advice is to experiment with two or more brands simultaneously so as to keep climate and duration variables constant. (It's simple on a resto-rod with front and rear bumpers, bumper guards, etc.) Either way—wax or squirt—keeping the bumpers clean is important in fighting corrosion.

A SPRINKLE OF STARCH

Heavy-handed use of a power buffer leaves swirl marks in black paint. Even manual waxing may leave streak marks if the paint is oxidized, as the wax penetrates unevenly. Try this time-honored method for removing such distractions, or even an excessive coat of wax. Lightly sprinkle corn starch over the surface while the wax is drying. Follow that up with a very gentle hand polish. It works well on some finishes, and only so-so on others.

PAINT ON THE RUN

Speaking of paint oxidation, those of you who've chosen red or blue for your car are probably aware that they are known as "fugitive colors" and tend to oxidize more rapidly than most others. (If you're not sure if your paint is oxidizing, just wipe the car down with a clean white rag. If you see color, you're in trouble.) We've heard nothing but good about Meguiar's four-part deoxidizing system.

Another automotive rehabilitation product, but for interiors, is called Color Bond. It is manufactured by BrynDana International (P.O. Box 1233, Havertown, PA 19083, 610/789-83031). Color Bond is a process for the repair of minor blemishes as well as the restoration and/or color change of leather and vinyl. It is worthy of your on-site evaluation.

OUTASIGHT

Street rodders like the clean look, and often find the radio antenna obtrusive. Dakota Digital, (3412 Hovland Dr., Sioux Falls, SD 57107, 800/852-3228) catalogs an electronic AM/FM antenna (ANT-1000). It is designed to receive signals at least as well as the conventional rod antenna. Only 3.5-inches long and 0.75-inch thick, it can be mounted anywhere in the car with its nine-foot cable making the need to drill holes in that precious exterior sheet metal unnecessary.

One of the many neat products I've run across on the well stocked shelves of my favorite racing emporium (Speed Shop U.S.A., 2365 W. Whittier Blvd.; La Habra, CA 90631) is Wipe Clean Water Spot Remover. The name says it all, and it really works.

Chapter 12

MISCELLANEOUS TIPS & TECHNIQUES

Street rodders sometimes toss the phrase "nuts and bolts" around loosely in casual conversation. Unfortunately, I'm sorry to say, real nuts and bolts are sometimes taken lightly. Most of us in this hobby for any significant length of time keep an ever-expanding collection of coffee cans. A felt-tip pen ID on a masking tape "label" is the primary means of differentiating those that contain flat washers from lock washers.

Now, I'm not throwing stones, I live in the same glass house. Only I make the labels for my coffee cans and plastic food jars on my computer. Nevertheless, I admit I am just as reluctant as the next guy to toss out old fasteners is. But I don't use them on my street rod.

VARIETIES

Conventional bolts (sometimes called cap screws) are used to clamp surfaces together into a rigid joint; nuts may or may not be necessary. The typical bolt used in street rod

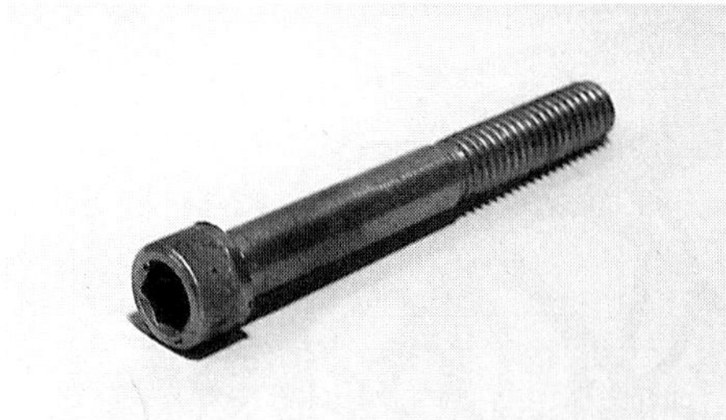

applications has a "hex head" with six flats.

The internally wrenched "socket-head bolts" (commonly called Allens after the original manufacturer) have a recess in the head with six flats.

And, occasionally, one will see aircraft-inspired 12-point heads on street rods built by ex-drag racers.

Studs are also great fasteners. Just remember that they are meant to be installed finger-tight as they need a little slack to avoid damaging the female threads in the work piece. (In other words, don't screw them all the way into the bottom of the hole.) By the way, larger studs sometimes have a small hex head and/or an Allen wrench pocket. Such features increase the cost, but when you need them . . .

WRENCH ROOM

Although 12-pointers and Allens are more expensive than hex-heads, both will go places where hex-head wrenches or sockets simply can't fit. The downside of both is the fact that there is less of a load bearing area under the respective heads. If you decide to upgrade your stock hex-headed caps screws, use hardened washers. (Soft washers are for soft jobs!)

Regardless of head design, the shaft extending from the bearing surface to the end of the bolt is divided into two parts: the unthreaded "grip length" (sometimes called the shank) and the "threaded length." Thread pitch refers to the number of threads per inch. There are two varieties in common use in the United States—National Coarse (NC) and National Fine (NF). For example, a half-inch diameter hex-head bolt will have either 13 threads (NC) or 20 threads (NF) per linear inch and will accept either a 1/2-inch x 13 or a 1/2-inch by 20 nut. (Yes, there are metric threads and several others used in the aerospace industry and professional racing, but with the possible exception of metrics, street rodders are not likely to use them.)

FINE THREADS

Fine threads were once considered weaker than coarse threads of the same metallurgical quality because they weren't as deep and didn't have the rigidity or the flank engagement of the coarse thread. Current thinking is that there is no measurable difference in fatigue resistance between the two. When vibration is a factor, however, fine threads are preferred.

SAFETY FIRST

However, when tapping an aluminum or cast-iron work piece, where the female thread is weaker that the male thread on a steel bolt, use coarse threaded bolts.

A GOOD FIT

There are several classifications as to how snugly the external thread of a bolt fits the nut or the internal thread of a work piece. They range from "loose fit" to "interference fit," but for most street rod applications, Class 2, a general-purpose fit suitable for non-critical applications at medium stress levels is appropriate. (Normally you don't have to ask for a "Class Two Fit," but if the sales clerk in an industrial hardware store asks, you'll know what to say.)

PROZAC TIME

The mention of fit leads to the subject of threaded fastener tightening. The primary purpose

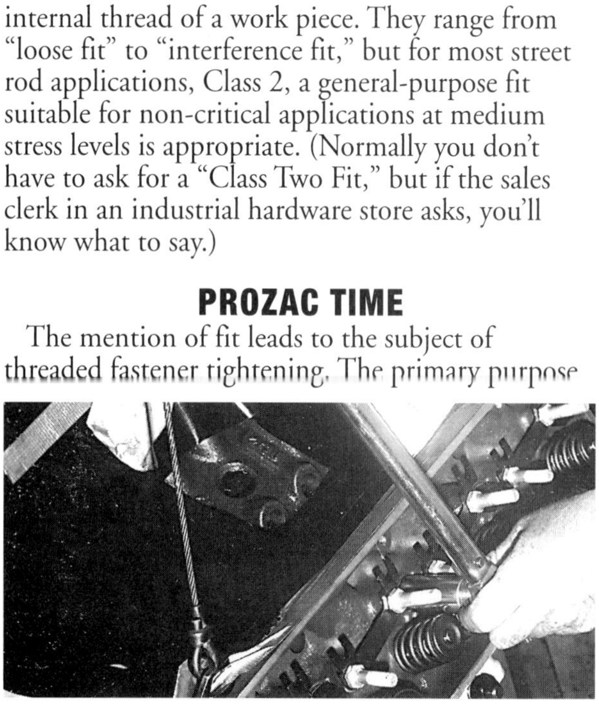

of a bolt or stud is to exert a clamping force. Clamping induces tension or "stress stretching" of the bolt being tightened. This stress remains until the bolt is loosened. Think of cylinder head bolts.

There is another stress, however. Shear is the lateral deformation caused by forces that produce an opposite but parallel sliding motion. If this is hard to visualize, think of flywheel bolts. Unlike tension stress (which remains constant as long the head is bolted down) shear stress is only induced when the flywheel spins.

By the way, you've probably noticed that flywheel bolts look different from most bolts. They are called "place bolts" and are easily identified by their slotted heads.

QUALITY COUNTS

The most important topic of all in any general discussion of bolts is the quality of those ubiquitous fasteners. Most of the bolts sold in discount home improvement stores should never show up in any critical street rod application. Blame it on the cost-consciousness of wholesale mass marketers who import inferior products from Third World countries or the callous disregard for safety by out-and-out counterfeiters. No less an end user than NASA had to disassemble one of the early space labs to remove counterfeit and defective fasteners. Simply put, you just can't be sure of what you are getting, due to unreliable package designation at such resources.

RELIABLE RESOURCES

Unfortunately, inferior bolts coming from overseas have been known to have head markings that are identical to those designated by the SAE for Grades 5 and 8! What's a prudent rod builder to do when he wants to install high quality fasteners in his suspension and steering systems and drive train?

Well, if you really want sleep-easy quality nuts and bolts, the three most reliable sources are machinery supply stores, industrial hardware stores, and general aviation supply houses. The latter are usually located near small airports that cater to the needs of private aircraft maintenance folk.

GOOD GRADES

There are several bolt standardizing systems—Military Specification (MS), National Aerospace Standard (NAS) and Society of Automotive Engineers (SAE). The most practical for street rod applications in terms of cost and availability are two SAE grades.

Grade 8 (five line) is recommended for critical applications and Grade 5 (three line) is adequate for most everything else.

Well, you ask, what about nuts? Good question. The problem is there is no hard and fast rule about nut grade marking. For as long as I can remember, I thought the markings on the ARP cylinder head stud nut (on the right) indicated a Grade 8 certification, and long ago I was told by a knowledgeable friend that the marked face should be visible when the head was installed. I never questioned either assumption until the day I counted the individual markings. There were nine slashes! Now, my dad was fond of saying "When in doubt, don't ask the monkey, ask the organ grinder." In this case, the organ grinder was ARP, and I was told by the Tech

Desk the markings were for company identification, and it didn't matter which side was up. But there's more.

Nut and bolt grades must match. Makes sense, right? Yet, the nut must be slightly softer than the bolt so that its threads can flow to fit the threads of the bolt. Hmmm. Quoting from one of the references forwarded to me by ARP, "Manufacturers use several different nut markings to denote grade identification. Some are marked on the top and others are marked on the sides. Grade 8 nuts are always marked, but grade 5 nuts sometimes are not." Well, if you are scratching your head as I am, here is my best advice. When nuts are needed for engine building, go right to the ARP catalog and order exactly what you need for each individual application. When nuts (and bolts) are needed for other than engine service, go to your favorite industrial hardware outlet, not the closest Builder's Buddy, and order your grade 5 or grade 8 nuts and bolts in marked boxes of a 100 count. Do not mix them up, and keep them accurately labeled.

THE HARDEST NUT TO CRACK

Here's the rule: For maximum safety and performance, a nut should be discarded once it is removed. I know what you're thinking . . . "Yeah sure, lug nuts too? I'll go broke in no time!" According to one of my references (Bowman Distribution) a nut acts much like the coils of a spring when an assembly is tightened. Each thread in the nut will carry progressively less load. And the first time a nut is reused, 20 percent of the tension is lost. On the second reuse, 33 percent of the tension is lost. In short, a nut conforms to the thread of the bolt, so it loses more holding ability each time it is reused. I would be in remiss to even hint you can safely fudge. Let your conscience and common sense be your guide.

DECENT EXPOSURE

Sometimes, you just can't seem to find the right length for your application. What you have is a little too short for the job, or a little too long for those of us who like to be neat and tidy. The rule of thumb for length in cases where a nut is used is to leave three exposed threads beyond the nut.

It's more critical when the bolt is inserted directly into the work piece. Use a depth gauge to determine the maximum distance you can run it down, and shorten the bolt by three threads before it bottoms out. Yes, it is permissible to shorten a bolt; you can hacksaw it or use a thin blade on a pneumatic die grinder. Taper the cut gently with a fine file.

THINK AHEAD

My old friend Bob Beatty taught me many years ago, that everything that goes on or into a street rod has to be removed sooner or later. Always keep that in mind when building the car, and later on when fabricating brackets and installing accessories. Along this line of thought, always keep a jar of anti-seize lubricant on your workbench.

DON'T LEAVE HOME WITHOUT IT

How many times have we all been fooled by "eyeball" matching and brought the wrong gasket or part home? Just because it looks right doesn't mean it fits right. Bring a sketch or a tracing or something with you to make an on-site verification for even the most routine purchase at the parts house or automotive dismantlers.

MARKET FORCES

Speaking of automotive dismantlers, remember that the operators of those cornucopias of potential street rod components, read car mags as well as you and other rod builders. If you're certain you need some newly touted low buck part, buy it before the demand rises.

A WEIGHTY SUBJECT

Every street rodder should know exactly how much his car weighs. The best place to find out is at a public scale. First get the total weight, and then weigh the car with only the front wheels on the scale. That done, drive it forward enough to get the weight on the rear wheels.

To determine the percentage of the weight on the front wheels, divide the front weight by the total car weight and multiply the answer by 100. To get the percentage on the rear, divide the rear weight by the total weight, and multiply that number by 100.

DONOR CARS

Newcomers looking for an engine and transmission often think first of late 1960s or early 1970s Mustangs and Camaros, largely because of their mystique. It should be obvious that the cars themselves are very much sought after for street machine projects and that means premium prices for more or less identical parts that can be found in substantially less popular and cheaper full-size passenger cars. Although you don't want to use 35-year-old expendables such as motor or trans mounts, nearly everything else can be refurbished.

THERE'S A BUT

The only exception to this axiom is the rearend. Full-size passenger car rears are usually several inches too wide for a comfortable fit under most rod material.

HELPING HAND TIPS

No matter how dexterous you may be, there will always be a time when you need some outside assistance. Here are a couple of tips when working in the tight quarters that exist on every street rod.

EVERYTHING IS BETTER WITH GREASE ON IT

If you have a straight shot at a bolt that needs a nut, but you can't get your hand in there, a drop of oil on the threads and a bit of axle grease on and in the socket pocket will hold a nut captive long enough to get it started.

Conversely, even if you can get a socket or wrench on the head of the bolt, there's still only enough room to slip an open end wrench in to hold the nut in place, a piece of masking tape will hold the nut and lock washer long enough to get 'em started while you gently tighten the bolt.

VIRGIN HARDWARE

Faced with installing brand new nuts and bolts in a very tight space? Lubricating and initially mating them at the bench makes the on-site installation easier.

CUSTOM-MADE TOOLS

Street rodding folklore has it that Henry Ford made his own wrenches before he built his first car. That may be a bit extreme nowadays, but if you have a recurrent task in a semi-inaccessible place, pick up some good used wrenches at the swap meet and modify them to your liking.

I not only had to reshape one of my distributor wrenches to work around the blower belt on the

race car, I also had to insulate the handle with a piece of heater hose to ward off those once-in-awhile disconcerting electric shocks. If you run into such a problem, you'll know what to do.

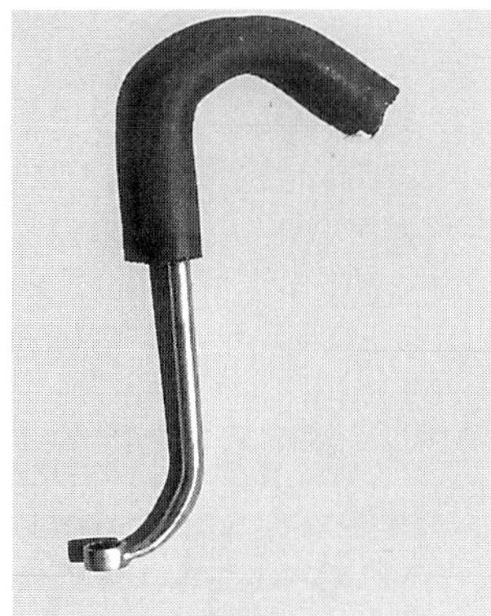

A DROP OF PREVENTION

I suspect you've noticed street rods have a tendency to shake, rattle and roll! Put a dab of thread adhesive on every fastener. In the same vein, some of you may remember that old cartoon of a bolt chasing a nut with the latter yelling "not without a washer!"

Teflon tape works great for sealing threaded connections, but in critical applications such as fuel lines the liquid variety eliminates any chance of debris.

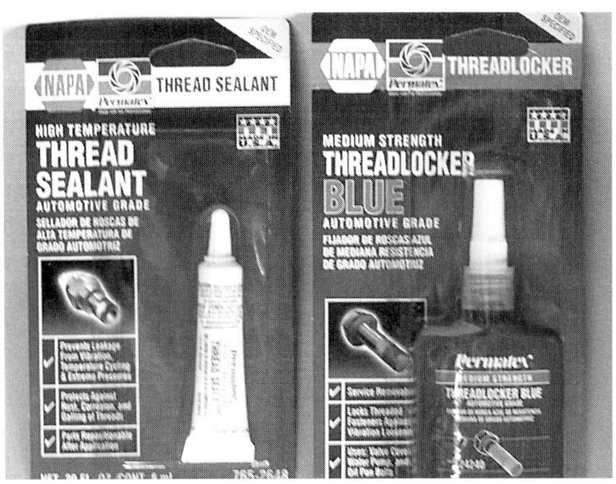

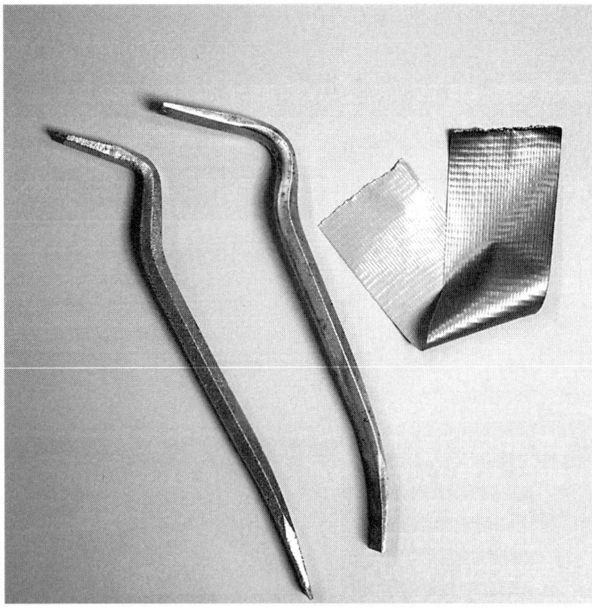

FOR RESTO-RODDERS ONLY

Do you run reproduction hub caps on painted steel wheels? An old brake spoon with its sharp or rough edges smoothed out makes a dandy hub cap removal tool. Don't forget to wrap a bit of electrical or tiger tape around the hump to prevent scratching the wheel.

SPECIAL LUG NUTS

For rods equipped with alloy wheels that use special lug nuts attach a set of stock lug nuts to the stock-wheeled spare you carry. An 11-inch nylon wire tie through the stock lug nuts and one of the holes in the spare's rim keeps 'em where you want 'em. Just make sure you carry a lug wrench (or wrenches) that fit both the stock lug nuts and the special ones.

ATLAS LEVERAGE

Sometimes you simply need a lot of leverage to break loose lug nuts previously applied with more torque than your half-inch impact wrench can deliver. And, that big lug wrench that always worked before sometimes bends when you put some muscle to it! Pick the most useless socket on the lug wrench; cut off just enough to slip your floor jack handle over and up the shank. (Don't laugh; I actually had to do this to remove the factory-installed wheels on my late model pickup!)

DRIVESHAFT STORAGE

If you've had your new driveshaft chopped and fitted perfectly to your drivetrain installation, but it's going to be a while before you actually stuff it in, prevent surface rusting of the bare metal components with Cosmoline. It's the best stuff I've used for this type of protection. Just be sure to clean the yoke carefully to avoid automatic transmission contamination when the 'shaft is eventually installed.

SIZING SANDPAPER

Every street rodder has several sanding blocks, electric sanders, and even pneumatic sanders. Of course, pre-cut sandpaper is available, but it's never there when you need a particular size. A fine-tooth hacksaw blade held firmly and squarely on the sheet allows you to make a clean tear. By the way, a strip or two of duct tape on the smooth side of ordinary sandpaper will add the strength needed for a power sander.

A DUST BUSTER

You spend your entire Saturday washing and waxing your car for the big parking lot show on Sunday only to have to spend another half hour dusting it down after you get there. You simply attracted a lot of dust thanks to the static electricity generated from the friction between the car and the atmosphere on your way to the show. Attach a length of chain to a discrete part of the chassis just long enough to drag the ground. Your homemade "static strap" will discharge the accumulated static electricity. You can take it off when you get to the show.

T-BUCKET GO PEDALS

Petite cable-operated accelerator pedals that are perfect for the smallish cockpits of T-buckets and channeled roadsters can be found in several subcompact models languishing in the automotive dismantler's back lot. Most are easily mounted to the firewall.

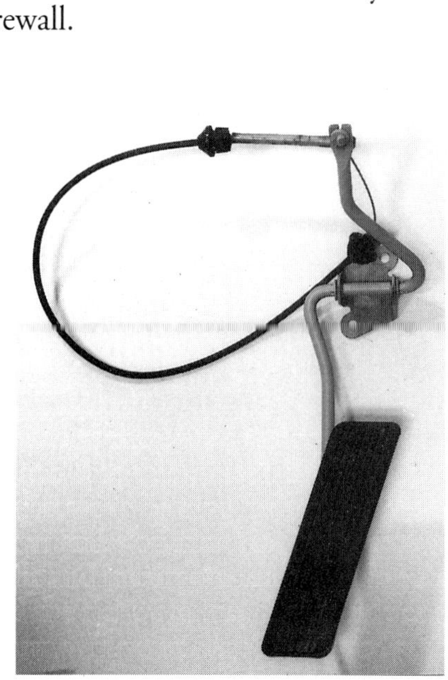

A QUICKIE DRILL DEPTH GAUGE

When you have to drill a hole to a specific depth with a hand drill motor, wrap a piece of masking tape around the drill bit corresponding to the depth desired.

BUDGETARY RESTRICTIONS?

Can't make up your mind between an oxy-acetylene welding set and a wire-feed welder? Go for the gas outfit first. It is more versatile.

Got air tools . . . but no compressor? Rent a bottle of carbon dioxide from the welding shop in the meantime.

DOWN TIME PREPARATION

Street rods, by the very nature of their owners, are always in a "planning stage" even when they are operable. The street rodder who is satisfied with his car's performance or appearance longer than six months is, if not a rare bird, at least atypical. His car is his "tinker toy" in the true sense of the term, and that means some kind of modification is always just around the corner. When it's going to be out of service for awhile, consider each of the following tips on an "as needed basis" depending on what the future holds for you.

Less Than a Year

Lube & Fuel Up—First lubricate the running gear, and then fill the gas tank to reduce condensation. Wash and wax the body. When you pull the car into the garage, change the oil and filter, and check the coolant level. Run the engine until it normalizes (180-degrees) and fully charge the battery.

Inflate the tires to 35–40 psi and chock the wheels. Close all the windows and disconnect the battery cables. Do not set the parking brake, however. Now you can drape that nice new cover you recently bought over the car.

Valve springs can lose some of their tension when they are compressed for months at a time. To avoid this, fire the engine and bring it to 180 degrees every six weeks or so. Pull and clean the plugs when the engine has cooled. Otherwise, they will "load up" with long term idling.

More Than A Year

Oil the Carb—Fire up the engine, and bring it up to operating temperature. As it idles along, slowly pour 10W detergent-free motor oil into carb. When it stumbles, shut it down. Remove the carburetor and carefully drain the remaining gas out of the bowl. Replace and adequately cover it. Then pull each sparkplug while the engine is still hot, and squirt a bit of 10W into the combustion chamber. Reinstall the plug.

Release the tension on all the springs that are holding valves open. Note the relevant cylinders on a strip of masking tape and attach the tape to the respective rocker covers.

Drain the Gas—Completely drain the gasoline out of the tank. And, if you didn't do it when you built the car, now is as good a time as any to install an easily accessible petcock in the fuel line to facilitate all subsequent tank drains. (You'll realize why this is an advantage while you're under your car undoing the fuel line!) Remove and store the battery in a well-ventilated area.

Roll On—There's one school of thought that recommends the car be jacked up enough to take the weight off the tires. Another resource contends that tires deteriorate more rapidly when they're unloaded. Inasmuch as I'm a believer in keeping most everything on wheels in a small shop for obvious reasons, I believe you should just leave the car on the ground with the tires on.

Apply a protective coat of paste wax on all exposed chrome plating, but if you live in a high humidity part of the country, it wouldn't hurt to put a coat of wax on all chrome, period. Do not buff out the wax.

RETURNING THE CAR TO SERVICE

Check all fluid levels, belt tensions and tire inflation. Clean and reconnect the battery cables, then put a slow overnight charge on the battery.

Gaskets and seals dry out over time, so check for any leaks—coolant, engine oil, transmission fluid, differential lube, brake fluid and gasoline before you fire up.

With all systems "Go," remove the spark plugs and squirt in an ounce or two of oil to help loosen any sticky rings. Readjust all valves previously loosened. Crank the engine over for 10-15 seconds to pre-lube it and expel any fluid accumulation in the cylinders. This will also replenish the fuel level in the carburetor.

Replace the spark plugs and start the engine. While it's warming up to operating temperature, monitor all the gauges. Let the engine run for about 15 minutes at 1,500 rpm and then change the oil and filter. You should also replace those carbon-fouled "warm-up" sparkplugs with new or freshly cleaned plugs. Check the valve lash on all cylinders.

Roll Out

When you're at last ready to back out of the garage, don't get too eager. Cruise around the neighborhood slowly for three or four miles to ensure lubricant circulation and drivetrain seal softening.

Apply the brakes often, but very cautiously, not only to check 'em out, but also to remove any potential oxidation on the brake drums or rotors. If the car pulls to one side or the other during these slow speed brake trials, inspect the wheel cylinders immediately upon your return home. It's very possible a piston "froze up."

Even if everything appears top notch, recheck all fluid levels and belt tensions, all the while keeping an eye open for delayed arrival leaks. That done, you can now wash and wax your baby.

ID NUMBERS

The combination of a light stamping and corrosion can make the frame serial numbers on 1935–48 Ford passenger cars and light trucks hard to find. They are located on top of the rail on the driver's side about 20 inches from the firewall.

Want to understand what the engine ID on the back of your early Chevy six means? Patrick's Antique Cars and Trucks has a (proprietary) Chevy Six Swap flyer that offers this tidbit: On the block above the end of the starter will be an initial (A-L) followed by either two or three numbers. The initial represents the month; the number is the day and year the block was cast.

TEE FOR SIX

Those bundled Allen wrenches are great. Sometimes, however, the longer "T-handle" singletons are critical. Old eyes sometimes have trouble "reading" the sizes; it helps if they are

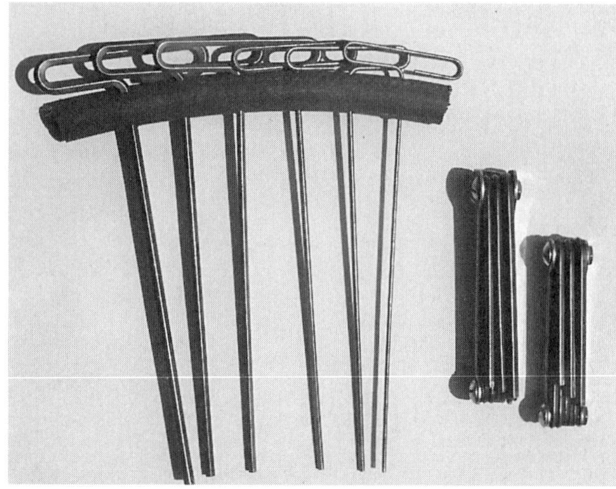

organized by ascending order in a perforated length of heater hose. That way they can be quickly removed from the tool chest drawer en masse and taken to the car.

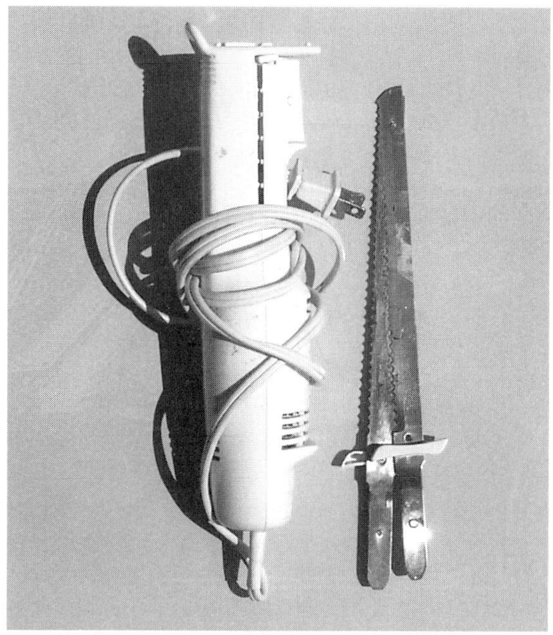

A KNIFE OF YOUR OWN

Don't go sneaking the electric carving knife out of the kitchen when your wife isn't looking. Stop by the discount store for your own; they are invaluable when you want to cut or shape a piece of foam rubber.

DOWN TO THE LAST DROP

Since gas prices have risen, street rodders have had to consider a new approach to their hobby—how to squeeze more mileage out of a gallon without sacrificing performance.

As this is written, the price of gas is on every street rodder's mind. Not that any one of us would

drive our rods any fewer miles than we did a year or two ago, our voices just join the chorus of complaints. Today, most street rods are owned, maintained and driven as a hobby. They are not the multi-purpose, transportation cars they were beginning in the Depression of the 1930s and on into the 1970s. Even though the truly utilitarian value of the street rod may have decreased, a measure of practicality is still worthwhile. So, in the spirit of "waste not, want not," here are a few tips on how to squeeze out better gas mileage without sacrificing the essence of street rodding—performance motoring in a modified vintage automobile.

Next to basic aerodynamics, the most important factor in gas economy is vehicle weight. Although no street rodder is comfortable without a complement of hand tools nearby, don't carry your entire toolbox on board. A few wrenches, a couple of screwdrivers, a pair of pliers and a flashlight are all that is needed.

Remember that when it comes to performance motoring, the objective is not to save money, it's to save gas. That's why it's wise to check the timing and valve lash (if you're running mechanical lifters) once a year or every 12,000 miles, whichever occurs first as they say. It's also a good idea to take a look at your carb linkage periodically. Make sure the butterfly is exactly perpendicular and the choke (mechanical or electrical) is opening fully.

It may be too late, but if you are still in the planning stage, keep in mind that although big displacement engines are often great fun, most of the time they're only extra poundage.

Gas mileage improves when engine speed (relative to road speed) is lowered because internal friction in the power train is reduced.

Modern automatic transmissions with overdrive deliver darn good gas mileage, but manual boxes still have less internal friction and slippage. The fun-factor of the latter, however, can negate any mileage gains. Prudent driving habits must always prevail.

The final drive ratio between road speed and engine speed in high gear is an important factor in the search for fuel economy. It includes the transmission in high gear (1:1 or about 0.65:1 with overdrive), the rearend gear ratio and the overall diameter of the rear tires. Stay away from those 4.11:1 gears.

SWITCH-HIT DRIVING

Most of us are switch-hit drivers; we go from driving the family car with its automatic transmission on a regular basis to periodically driving our street rod. If it's a stick-shifted transmission, some of you will probably find your feet in the wrong place at the wrong time.

The typical problem with a manual tranny in a street rod is giving it too much pedal when taking off from a dead stop. It takes a concerted mental effort to synchronize the proper throttle and clutch release required to smoothly launch the car from a dead stop every time. The solution to both problems? Simple, drive your street rod more often!

When you come to a stop signal, shift into neutral and release the clutch. That saves wear and tear on the disc. When the signal changes, drop into first gear, and leave when cross-traffic permits. It is not necessary to "speed shift," but shift quickly from gear to gear. And don't make the common mistake of "blipping" the engine while momentarily in neutral. There is no up side; you're only wasting gas.

The point is to get into high gear as soon as possible and to stay there for as long as possible. Playing road racer and messing around in the lower gears may be fun, but when engine speed is out of proportion to car speed, fuel consumption increases. Once in high gear, your left foot should be somewhere comfortable and ready. Inasmuch as the cockpit of a pre-WWII car is smallish, relocating a floor-mounted hi/low beam switch is often a wise move.

It is easy to forget the truism that all automotive energy is derived from burning gas. Surely you didn't think that electrical accessories were cost-free just because they are battery powered? What keeps the battery charged? For instance, the extra electrical load demanded by the high beams can decrease mileage almost a half mile per gallon!

The recipe for a basic, no-frills 300–350 cubic inch street rod engine with enough beans to handle the road, yet is easy on the wallet at the gas station

is a short-duration, mild street cam, a street-oriented manifold, a 500–550 cfm carb with vacuum-operated secondaries, and "shortie" tubular headers with low restriction mufflers. It will typically show a fuel economy increase of five to fifteen percent over a comparable displacement engine where the components were selected on the so-called "high performance" basis.

Not always sure about the phrase "street oriented" when it comes to catalog jargon? Substitute "RV." Components such as these are doubly effective when they are backed up with a lower numerical rearend ratio (3.00:1 – 3.50:1). Low-speed torque is the key to fuel consumption efficiency and enjoyable performance motoring.

There's no doubt it's a bit of work, but switching to a lower numerical ring and pinion gear set can show significant gains in your mileage calculations. Numerically lower gear ratios not only permit the engine to operate at lower rpm; reducing rpm cuts the internal friction in the power train. Since friction increases with the square of the speed, a seemingly minor reduction in rpm can eliminate a substantial amount of friction.

CHECK THE MPG

Sometimes street rodders have an aversion to knowing the exact mileage their pride and joy delivers. I suspect it is because of a subconscious guilt complex. However, what you don't know can obviously hurt you.

Start with a full tank, fill it to a reasonable level, check odometer and note the mileage. Drive until you've used a least half a tank; the more you use before you refill, the smaller the error. It's not easy with most street rods, but try to refill to the same level, and do it two or three times to obtain an accurate picture of the car's fuel consumption. Do this little exercise every few months, and any significant changes will tell you something isn't kosher.

For those of you who live in Yankee states, warming up the motor is a necessary evil. Just don't overdo it. The best rule to follow is to warm the engine for just a short time until normal oil pressure is registering on the gauge and then drive gently until the temperature of the engine normalizes at 180 degrees.

BUILD A DATABASE

When you see something in an automotive magazine that may be of interest to you down the road, be it an article or the address of a supplier or resource, immediately file it in your computer or on a 3- x 5-inch index card. Like old parts, you never know when you will need old information.

SPECIAL TOOLS

Snap-On's Nut Splitter

There's nothing quite like working on old cars. The simplicity of design so often found in veteran automobiles has long charmed mechanically minded Americans. Yet the very fact of advanced age engenders problems not met in less seasoned vehicles. And probably the most exasperating hassle of all is rust. It is the first nemesis the street rodder encounters as he begins his project. It eats away at valued tin, falls into unwary eyes and turns ordinary nuts and bolts into recipients of profanity. But if street rodding is your game and you plan to pursue it for any length of time, you must face up to the inevitable fact that rust is always going to be the fly in your ointment.

Most street rodders aren't professional mechanics and consequently don't have the knowledge of, or access to the wide range of equipment necessary to earn a living at automotive maintenance. Yet there are a few relatively obscure hand tools that the hobbyist will find quite handy. One of them is the nut splitter. Like you, I haven't always owned one, and I've used my share of broken hacksaw blades clamped in Vise Grips to chew away at rusted and rounded nuts and bolts. I've got the scarred knuckles to prove it.

One afternoon long ago, I needed to pull a front fender off one of my parts cars. With more optimism than experience should have permitted, I crawled under and attacked the reluctant fasteners with wrench and ratchet. In no time flat I was spinning most of the bolts. My first impulse was to torch the fender off and damn the inner panel and chassis. But then I remembered Bob Beatty and his fascinating truckload of Snap-on goodies. It wasn't long before I had my own Nut Splitter (NC52). Here's how the thing works.

When you can't torch it off, crack it with the nut splitter. This simple but effective tool encircles nuts up to 3/4-inch, and makes quick work of rusted or frozen nuts on an inaccessible bolt. The stout forged body can be turned to the best position for splitting action and held with the bar. The pressure screw has a 3/4-inch hex head for socket or wrench application. The screw is tightened until the hardened cutting chisel cracks the nut.

Badly rusted nuts may require an opposing cut to finish the job, so make a second bite on the other side. A box-end wrench works when a ratchet won't fit. The split halves of the nut should pop off, but following the rule of the Intransigence of Inanimate Objects, sometimes they don't and have to be pried off. Once freed, the bolt can be driven out. The operation takes only a few minutes, but can save an hour.

The Impackdriver

OK, the nut splitter can be a big help when a decades old nut is accessible. But what about a rusted screw or bolt when the nut isn't accessible? Well, another handy tool is available, and it is called, appropriately enough, the Impakdriver.

It is basically a stout handle with a 3/8-inch drive stud on the end. It is spring loaded and will take any 3/8-inch drive socket or accessory. There isn't any adjustment, yet it will loosen or tighten a screw or bolt. Simply attach the appropriate socket or screwdriver bit to the handle, line it up and smack it with a two-pound hammer. At the same time give the handle a slight twist in the direction you want to go. The spring action follows your lead.

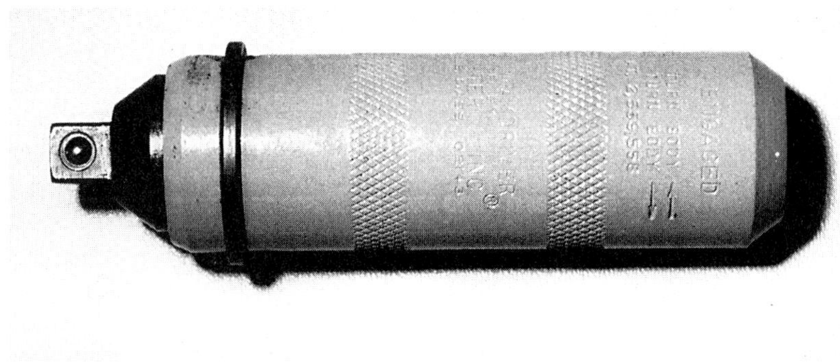

The Impakdriver helps remove nuts, bolts, and screws that are rusted, frozen, or corroded. The user fits the tool with the proper bit or socket and selects either the tightening or loosening function. The Lowell Corporation of Worcester, MA manufactures it.

The user then engages the object to be loosened or tightened and strikes the head of the tool sharply with a hammer. The Impakdriver converts the hammer blow into torque that loosens or tightens stubborn fasteners.

Judge the intensity of your blow by the apparent degree of corrosion retaining the screw. In general, you should start with light taps and increase the force as necessary. We're told the spring mechanism multiplies the hammer blow up to about 90 foot pounds of torque!

There is one qualification, however. If the screwdriver slot or bolt head hexes are badly weakened by rust they will crumble. Can't help you there, you'll just have to do it the old-fashioned way—they will have to be drilled out. Nevertheless, the Impakdriver will do your bidding in many cases.

The Impakdriver brings to mind the general topic of black vs. chrome-plated sockets. Not all black sockets are designed to be used with pneumatic or concussion tools as is commonly supposed. Those intended for use with such tools have a thicker wall than chrome-plated sockets and the alloy used and heat treatment are different. True impact sockets are actually softer than standard-duty sockets so they can survive the shock of a pneumatic impact wrench.

With regard to brand-name, high-quality, American-made black sockets, if they have a wall thickness about the same as chrome-plated sockets, it is likely they are intended for use in an industrial environment. And the reason has nothing to do with heavy-duty! It's the "crow" syndrome." No kidding, crows and sticky-fingered co-workers like to pick up shiny sockets and wrenches. Manufacturers determined tool losses could be reduced if black-finish tools were used. (I was reminded of the "crow syndrome" by a recent news article about crows stealing quarters out of a coin car wash change slot. Their hoard of quarters was found on the roof of the building. So watch out for the devil birds.

TECHNIQUES: PICKING THE "MUST HAVE" GAUGES

Most street rodders are fascinated by automotive instrumentation. Perhaps they fondly recall boyhood memories when they excitedly climbed aboard a fire engine for the first time and saw all those gauges. And why not? There is a certain amount of beauty in gauges. But make no mistake, be it a fire engine or a street rod, gauges are more than just a row of pretty faces. Utility and accuracy is what really counts. That string of dials stretched across the dash is the only way to keep abreast of what's going on under the hood and elsewhere as you motor along.

Now, I wouldn't go so far as to say that a street rod couldn't have too many gauges. I've taken my share of ribbing over the years because of the compass I keep in my coupe. But that's okay, the very first automotive gauge was a pointer fitted to the steering handle of Benz cars in the late 1800s to show the driver which way he was going!

There's never a wrong time to install a set of high quality gauges, but there is a best time—just before you fire up that fresh engine and take your car out for its maiden voyage. So if you are building a new rod, or just upgrading the automotive love of your life, take the first opportunity to latch on to a matched set, and as mentioned earlier, figure out where you are going to mount them.

The gauges in your dash are a major tool with which to analyze the performance of your engine. Presumably it was top notch when you installed it, but perfection started to slip away ever so stealthily as soon as you started driving it regularly. And it will be wearing out during the rest of its service life; it's a normal, constant process.

The gauges monitor the engine's journey through the years, and alert the driver to the little problems before they become big problems. In other words, they provide an efficiency baseline if you record the gauge readings under a variety of conditions for later reference.

The traditional street rod gauge is of the analog type; that is, it displays information to the driver by means of a pointer moving across a calibrated scale. The position of the pointer and the numbers on the scale reflect measurable quantities. Street rodders leaning a bit more to high tech than tradition often prefer the digital readout gauge, which displays its information directly in numbers. Nostalgia aside, there is some evidence that analog gauges are easier to read at a glance than digitals. When one is familiar with the position of the analog pointer in the gauge's "normal" range, he is instantaneously aware of the engine's condition.

There is another consideration when picking new gauges—mechanical or electric. Reliability-wise, neither has a significant advantage. Electric gauges, however, are far easier to install.

There are many different gauges rodders can buy. It is possible to make your dash look like the business office of a 747 if you like. However, there are the six basic instruments that you simply must have if you wish to know what's happening in your car at the moment and over time. Although some gauge manufacturers combine the functions in order to conserve space (a commodity in short supply on many street rod dashboards) individual gauges are easiest to monitor.

The Speedometer

It may not be the most important gauge on your dash (unless you're entering a radar zone), but no doubt it's the one you think of first when you think of gauges. And, the speedometer is the one you'll look at most frequently. Historically, one of the earliest (and certainly the most interesting) speedometers was the 1905 Cowley Recording Speed Indicator. It not only showed the speed at which the car was moving, but also the speeds reached over the preceding quarter mile. Just think, if you were caught in a police speed trap, your speedometer could prove your innocence. At least that's what the advertisements said.

For decades, street rodders stuck the basic black dial mechanical speedometer driven by a flexible cable from the transmission into their stock or custom dash with hardly a thought. Internally, the speed-indicating mechanism consists of a revolving

permanent magnet driven by a magnet. Stewart-Warner and others are still manufacturing it although the black dial has undergone a few changes through the years, some of them nostalgic. Aside from a rare cable failure, the only consideration over time was the need to change the plastic, color-coded transmission gear if the rodder decided to change rear tire diameter or back axle gear ratios.

The landscape changed when electronic speedometers appeared, the latter boasting "programmable" features that allowed "easy calibration with any tire size/axle ratio."

The Tachometer

The tach reads engine revolutions per minute. It is not an absolutely critical gauge, but it is useful. You can use it to adjust idle speed and to identify shift points for a manual transmission. It can also be valuable in a rod equipped with an automatic transmission. An increase in rpm is understandable when going uphill, but a steady cruising speed on a level highway should produce a steady rpm. Periodic increases in rpm could suggest internal automatic transmission slippage.

Although one sometimes sees a mechanically driven tachometer on a veteran sports car or street rod, today's standard is the self-contained electronic unit matched to ignition impulses, often with a "recall" reminiscent of the 1905 Cowley speedo. They are so accurate that even racecars use them in lieu of the one-time favorite mechanical unit produced by Jones of England.

Ammeters and Voltmeters

The ammeter indicates the output of the alternator in amperes. That is, how much electrical energy is flowing into or out of the battery, and whether the car's electrical equipment is consuming

more or less electricity than the alternator is producing.

Typically, the charge increases from idle to about 2,500 rpm. At higher engine speeds it should hold constant. A battery in good condition should show a charging rate of 20 amps right after the engine is started, and when the car reaches 25 to 35 mph. Within a few minutes, however, the charge rate should fall to 5 amps or less. If the charge rate decreases significantly while you are driving, something may be going wrong—a fan belt is loosening or worse—there is a failing rectified diode in the alternator. When you see an excessive discharge rate, you most likely have a short in the lighting, horn, ignition, or accessory circuits. Switching each system on and off usually isolates

the offender, but find it fast, or disconnect the battery. Fires often start that way.

An interesting sidebar to this discussion of automotive gauges is that the ammeter is not a precision instrument. The readout is an approximation; it is an average reading. Think about it, the ammeter cannot be too sensitive because of everything that is going on in the car's electrical system. If it were, the pointer would be hopping and bopping all over the place. It would be impossible to know if there was a problem.

The voltmeter (or battery condition gauge) is more laid back. All it has to do is monitor system voltage. It should indicate 12–13 volts when you turn the ignition on. When the engine is running, however, the voltmeter indicates the combined voltage of the alternator and the battery (i.e., the total operating system voltage) which should be from 13.3 to 15.2 volts.

Low battery voltage is nothing to be concerned about as long as it increases when engine rpm increases. Conversely, a high voltage reading may indicate a faulty voltage regulator. That's not good, an unprotected circuit could blow.

Now, which gauge, ammeter or voltmeter should you choose when dashboard space is limited? The consensus of opinion is the voltmeter. Every time you slide behind the wheel it will tell you what condition the battery is in. (Even before you turn the key, if you wire it that way!)

Coolant Temperature Gauge

Probably the most important gauge in the street rodder's frame of reference is the coolant temperature indicator. And yes, although they perform the same function, modern dash-mounted gauges are a little more sophisticated than those

long ago thermometers molded into handsome, sculptured radiator caps. Non-electrical temperature gauges incorporate a modified Bourdon tube design with a flexible conduit about 60 inches long. A probe on the intake manifold fitting dips into the engine coolant passage. Although they are not particularly fragile, care must be taken not to kink or damage the conduit.

Electrical temperature gauges are a lot easier to install than mechanical types. The sending unit too, has a probe that threads into the intake manifold coolant passage, but the sending unit is connected by wire to the dash unit with operational current provided through the ignition switch. The innards of the gauge are usually of the balancing coil-type.

Street rod cooling systems are pressurized and most rodders use glycol-based, anti-boil coolants so they can live with higher operating temperatures. The typical street rod engine "normalizes" in the 180- to 190-degree range, however. Although the temp gauge isn't ordinarily used to diagnose reasons for boil-over or coolant loss, an accurate gauge, constantly eyeballed, sure tells you when it is time to make an on-site examination of the usual suspects.

Oil Pressure Gauge

In the days before the widespread use of a mechanical oil pump, and some engines were lubricated by the "splash system," and an oil pressure gauge wasn't a big deal. Not so today. The second of the two indispensable instruments in a street rod is the oil pressure indicator. Mechanical gauges use a modified Bourdon tube into which oil is plumbed from a port in the oil galley; electric gauges have traditionally used the d'Arsonal

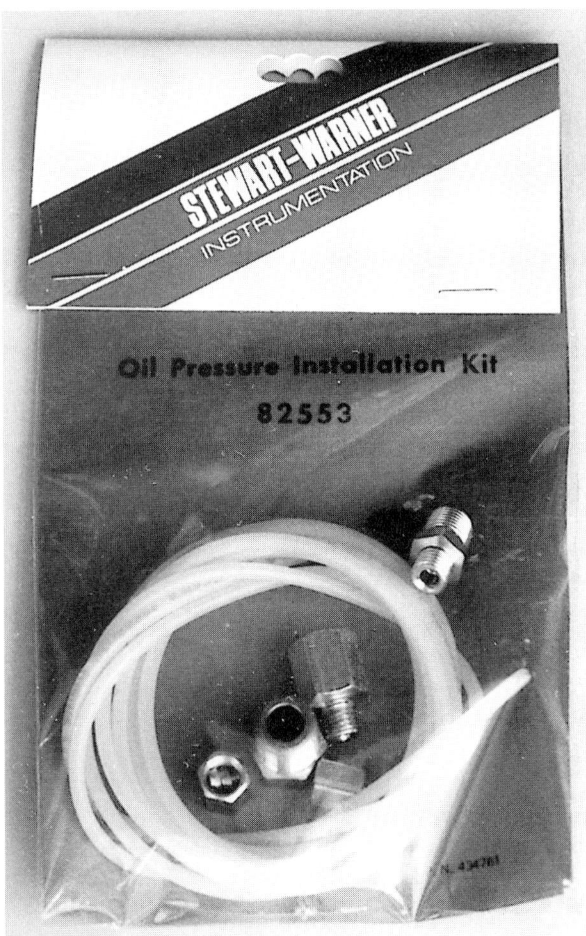

galvanometer movement which consists of an armature pivoting within the field of a powerful magnet coupled to a matched sending unit that encloses a diaphragm and a rheostat. A drop in oil pressure changes resistance and sends a signal to the dash unit.

Analyzing your engine's health during its service life is not unlike monitoring your own vital signs. However, as mentioned above, a baseline is needed first. So when the power plant is young and vigorous, record the normal oil pressure at idle then at 30 mph, and again at 60 mph. Naturally the engine should be warmed up and normalized, filled with fresh oil, and fitted with a new high quality filter. (The latter tips are also good reasons to use the same weight oil and brand of filter for subsequent testing.)

Expect some oil pressure changes as your engine breaks in, mellows out after 10,000 miles, and then grows old. If, however, you get an unusually low pressure reading someday, check the oil level first, and then look for loose or broken oil lines. Your coolant temperature gauge will tell you if you have an overheated engine. If the problem persists, and you haven't mistakenly put in the wrong viscosity oil, change the oil filter before you start looking for engine bearing failure. (I got fooled once; I was sweating bullets until I changed filters and the pressure "problem" disappeared!)

Fuel Level Gauge

Then there is the gas gauge, the constant harbinger of bad news these days. The tank sending unit has a rotating arm fitted with a brass float that varies the value of the current that flows to the dash gauge. When the tank is full, resistance in the sender is high, and the pointer so indicates. As fuel

is consumed, the float drops and so does sender resistance. Eventually, of course, the pointer armature is attracted toward "E." Sorry 'bout that painful fact.

One more thing . . . when installing gauges under the dash, a small piece of masking tape twisted around the eighth-inch plastic or copper tubing prevents those slippery little brass nuts or ferrules from sliding down out of reach.

TECHNIQUE: READING A VACUUM GAUGE

Finally, there is the vacuum gauge. To many street rodders, it is not a "must have" instrument, and when dash space is tight, it is left out. It is important to me, however, because of the way I've configured my coupe through the years. (I've sometimes run a 4V carb, and sometimes a 3x2V setup. I've also run an automatic, but on occasion, put my 4-speed Top Loader back in! And yes, I've been accused of not being able to make up my mind.) As a consequence, though, I'm a strong believer in the vacuum gauge. Here's why . . .

In a very real sense, the internal combustion engine is not just an "air pump" as you have been informed countless times. It is also a "vacuum pump." Vacuum, simply defined, is emptiness and it can be "produced" in a closed chamber by evacuating the air, fluid, etc. In an engine, intake manifold vacuum is developed (or maintained) every time a piston moves down its cylinder on the intake stroke. As the piston descends, it leaves a partial vacuum behind it. And, as you will recall from high school physics, nature abhors a vacuum. Consequently, atmosphere pressure outside the engine pushes air into the carburetor, mixes it with the vaporized gasoline, and sends the charge past the open intake valve into the combustion chamber and cylinder.

Intake manifold vacuum varies under different operating conditions. But more importantly, the manner in which it varies can be interpreted in an analysis of a wide range of engine defects. To know what's going on in your rod's powerplant at all times and to use this information as an effective way of enhancing your gas mileage merely requires the installation of a vacuum gauge, an instrument that can pay for itself within 1000 miles of driving at today's high fuel prices.

An automotive vacuum gauge is a relatively simple instrument. Essentially, it contains a balanced diaphragm that reacts when air is removed from its backside. (The removal of air is more abstractly known as the application of vacuum.) An indicating pointer geared to the diaphragm moves with it.

The scale of the gauge is usually divided into 30 "inches." This corresponds with the standard 30-inch column of mercury in a physics lab barometer at an atmospheric pressure of 15 psi.

Since vacuum gauge readings are calibrated for use at sea level, the observer must mentally compensate for those taken at higher altitudes. Such readings will drop one inch for each 1000 feet above sea level. Therefore, a reading of 20 inches vacuum at sea level would be only 19 inches at 1000 feet, 18 inches at 2000 feet, and so forth.

Vacuum gauge connections can be made at any convenient source of true intake manifold vacuum. The gauge must be installed ahead of any vacuum boosting device and ahead of any check valves in a power brake system. You should not use the carburetor as a source of vacuum as it can cause inaccurate gauge readings. However, most carbs have a plugged quarter- or eighth-inch pipe thread hole located in the base. This will work, but the preferable location is a tapped hole in the intake manifold itself.

Reading a Vacuum Gauge *(continued)*

Care should be taken in routing vacuum hose or plastic tubing. Avoid throttle linkages, and cowl vent handles, etc. Do not make any sharp bends, and where the line passes through the firewall, provide some protection against chaffing or laceration. Other than these precautions, installation is quite simple. Dashboard mounting, of course, is directed to a location that provides easy reading from a normal driving position.

Tune-up readings are taken after the engine has warmed up to the normal operating temperature and idling between 300 and 600 rpm. If everything in the engine is top notch, the pointer should be between 17 and 21 inches. Exactly where depends on your altitude, but the reading should be steady. When cracking the throttle on an engine in good running order, the pointer should drop to about two inches and jump back up to 25. Readings beyond this mean trouble and are covered in the accompanying diagrams.

Vacuum gauges, however, do much more than diagnose problems. They can be used to obtain the most economical operation of the engine while driving. There is a limit to the amount of fuel an engine can use effectively, and a prudent street rodder can observe this limit by restraining his right foot so that the gauge never reads below five inches. When it does, more gas is being fed into the engine than can be burned, and the surplus goes out the pipes or into the crankcase (thereby diluting the oil). In general, the higher you keep the pointer reading, the greater will be your mileage.

When the pointer drops below 9 or 10 inches under load conditions, it is an indication that you are "lugging" the engine. If your car has a manual tranny, or an older automatic, drop down a gear. Most late model automatics will do this for you.

Again, depending on the type and model of your transmission when you are braking against engine compression on a downgrade, the pointer may rise above 20 inches. Under these circumstances, a high mileage engine could suck oil up from the crankcase. Crack the throttle occasionally to get a lower reading.

Finally, I must point something out to those rodders who just have to install a high lift cam with lots of valve overlap. It will produce a lower and more erratic intake manifold vacuum reading. On the other hand, most mild street cams will deliver a reasonably true reading.

In summary, the vacuum gauge is a useful complement to the instrument cluster of any street rod. Hopefully your dash can accommodate one.

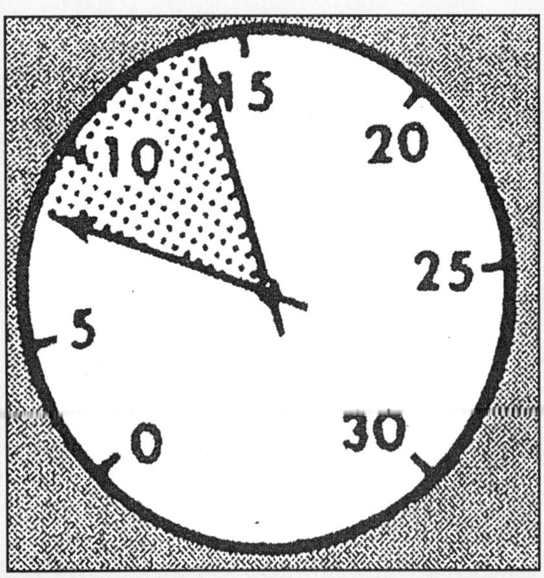

Incorrect Valve Timing: An appreciable variance in valve timing will produce a low steady reading between 8 and 14 inches.

Worn Rings: A reading lower than normal, but with a steady needle can indicate leakage around pistons due to stuck or worn rings, damaged pistons, or worn cylinder walls. When power is reduced, the engine will not "pull" full vacuum.

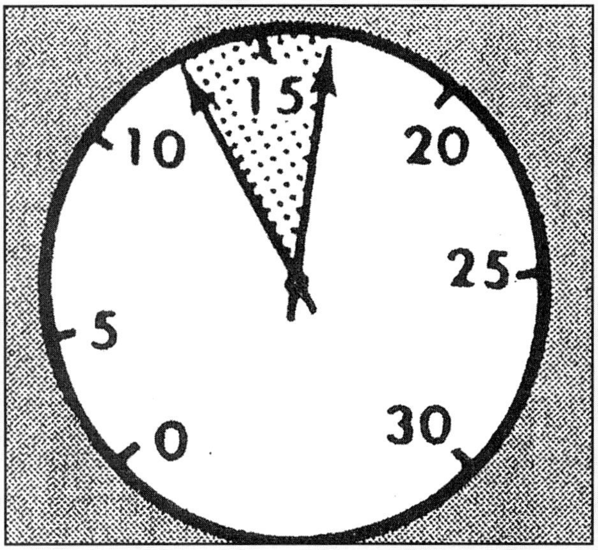

Incorrect Ignition Timing: If the ignition timing is off, the reading will be steady between 13 and 16 inches.

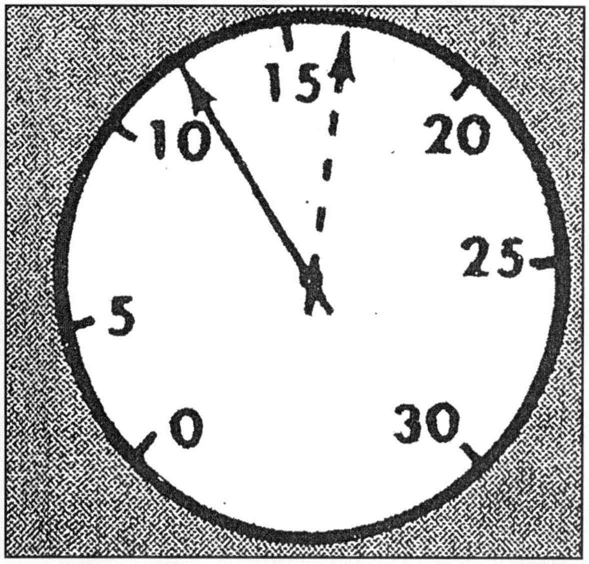

Poor Carburetor Adjustment: When the needle floats slowly between 12 and 16 inches, it will be necessary to adjust the idle mixture screw either richer or leaner so that the highest reading can be obtained with appreciable vibration. If you cannot correct the condition thusly you may have more serious trouble within the carburetor.

Leaky Carburetor or Manifold Gasket: A steady reading below 5 inches indicates a leaky intake manifold gasket or leaks around the throttle valve shaft.

Compression Leak Between Cylinders: When the needle drifts between 5 and 19 inches you've got a compression loss. If the needle vibrates while idling, but steadies as rpm is increased, chances are the intake valve guides are worn. Excess vibration at all speeds may mean just a leaky head gasket.

Plug Gaps Too Close: When the need drifts between 14 and 16 inches, pull your sparkplugs and check the gaps.

Sticking, Burning or Leaking Valves: When your valves are burnt or there is insufficient tappet clearance, the needle drops several inches each time a particular cylinder fires. If the needle drops irregularly, the valves are sticking irregularly.

TECHNIQUE: TAPS AND DIES

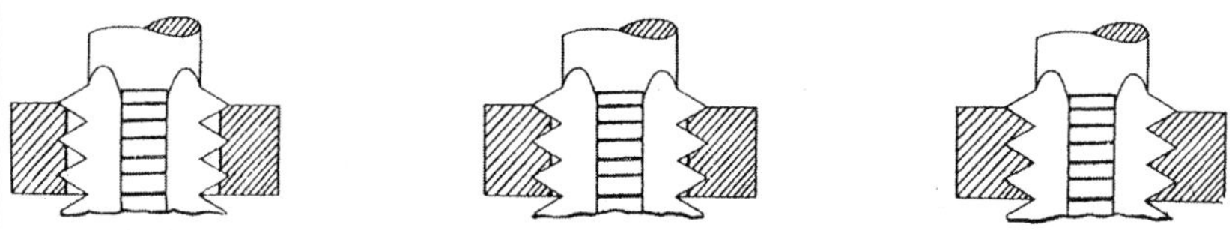

(A) 50% Thread Is Too Loose

(B) 74% Thread Is Best

(C) 100% Thread Is Too Tight

Indispensable is a heavy word, but few would argue the indispensability of a complete set of taps and dies in any street rodder's garage. Now, even though most may think they know how to work with them, there could be a few tips that have escaped your attention.

A tap is a tool used to cut material away to form internal (female) threads such as those in a nut. Besides cutting internal threads, however, damaged or rusty threads can be repaired with a tap.

The shank of the tap is ordinarily locked in a special handle called a tap and reamer wrench or an adjustable T-handle tap wrench. However, for those very tight spots where only a clean up is required, one can improvise and use a four-point socket or even an ordinary open-end wrench.

External threads such as those on a bolt are cut with a die. Round or hex-shaped dies are used in an adjustable die stock. Hex dies, however, are more desirable than round ones because they can be used with a wrench or socket in those inevitable hard-to-reach places. Dies can also be used to repair damaged threads.

Under no circumstance, however, should you die cut new threads in a bolt used in a suspension or steering component. Die cutting threads across the grain of the metal interrupts the flow of stress through the bolt and weakens it. (If you are wondering how bolts are manufactured, the threads are "rolled" on by wedging the blank between two or three rotating cylindrical dies, or a stationary and a reciprocating die.)

As shown in the illustration above, the effect of hole or bolt size on thread depth is important. For instance, if the hole is a little too large for the tap, the threads may be too shallow. (A) If the hole is too big, the thread will be too loose. (B) The recommended thread depth for general purposes is 75 percent. This provides near maximum strength with no danger of ruining the tool. (C) If the hole is a little too tight, however, the tap will be overloaded, a condition sure to shorten its service life. The same considerations hold true when threading a length of mild steel rod with a die.

Ordinarily, tap drills are larger than the root diameter of the tap and those in the Table will provide about 75 percent thread depth. Taps and dies, like all fine tools, should be cared for properly. This means keeping them clean and dry and returning them to the storage case when the job is done. Okay? Okay. Now, here's how to use 'em.

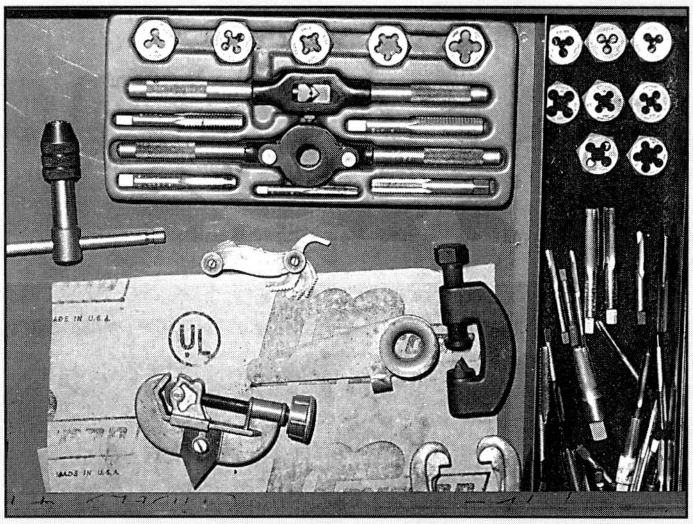

Basic tap and die sets are priced according to quality and completeness. The better brands and those that are more inclusive are sold in industrial hardware stores. Pieces can be bought separately—a plus factor since re-sharpening is not practical.

The first step with fresh work is to lay it out properly. The right hole size is a must. Using a drill bit that is too small doesn't mean a better thread—it only overloads the tap. After the hole is drilled, countersink it. When it is practical, use your drill press to start the tap square with the hole.

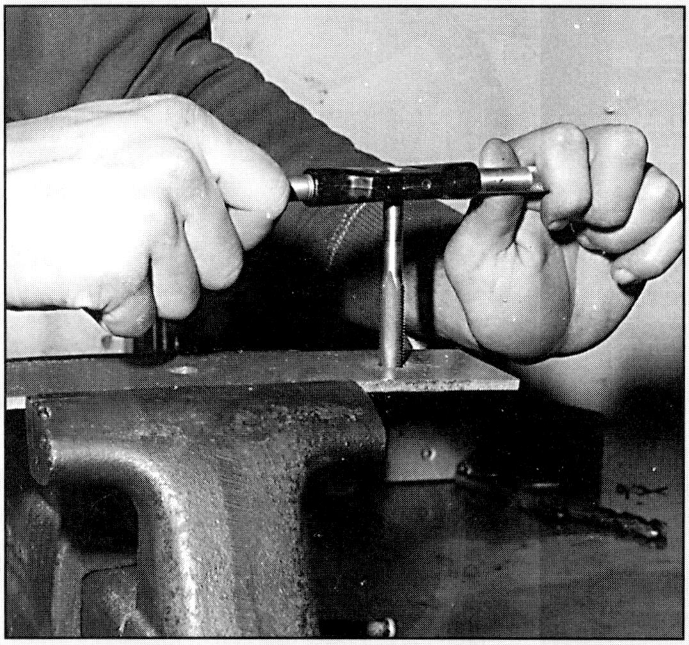

When starting the tap by hand, extra care must be taken to insure alignment. Freely lubricate the tap as it cuts into the metal.

As the tap starts to lead into the hole, back off every quarter revolution. This breaks the chips and releases the pressure on the tap. Continue until the required depth is reached.

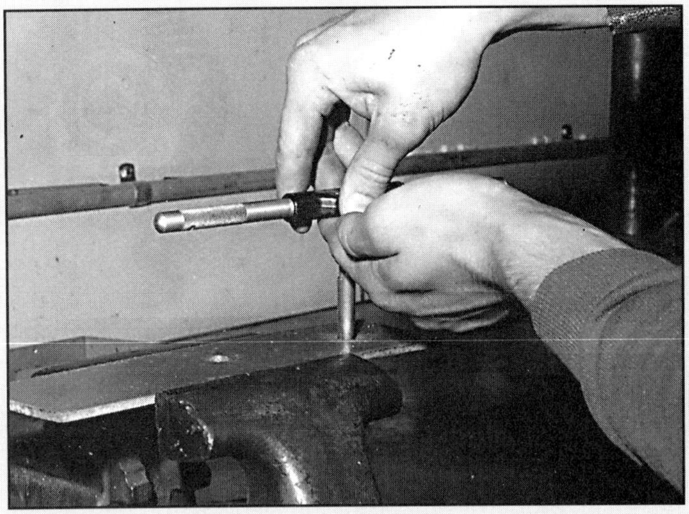

When tapping a blind hole, be certain to allow clearance at the bottom for chips and the tap chamfer. When removing the tap, support its weight as you near the top. Those first threads are easily damaged. By the way, taps and dies should be replaced at the first sign of dulling.

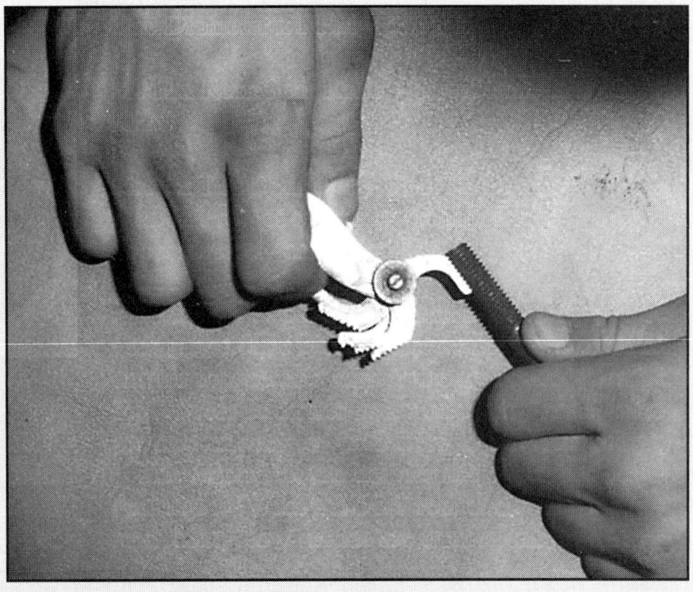

Unless you've got a great eyeball, the use of a thread gauge is highly recommended. The number on the blade indicates threads per inch.

When tapping a hole in a component where there is danger of metallic contamination, a daub of axle grease on the threads of the tap will retain the shavings for a while. Clean the tap frequently. By the way, keep dirt out of those little "tubs" of axle grease, white grease, etc. Cut an "X" in the plastic cover just large enough to poke a small brush through.

TAP/DRILL SIZES

Tap Size	Fractional Drill Size	Standard
4x40	3/32	NC
6x32	7/64	NC
8x32	9/64	NC
10x24	5/32	NC
10x32	5/32	NF
12x24	11/64	NC
1/4x20	13/64	NC
1/4x28	7/32	NF
5/16x18	17/64	NC
3/8x16	5/16	NC
3/8x24	21/64	NF
7/16x14	3/8	NF
7/16x20	25/64	NF
1/2x13	27/64	NC
1/2x20	29/64	NF
9/16x12	31/64	NC
9/16x18	31/64	NF
5/8x11	17/32	NC
5/8x18	37/64	NF
11/16x11	19/32	NC
11/16x16	5/8	NF
3/4x10	21/32	NC
3/4x16	11/16	NF
1/8 Pipe	5/16	
1/4 Pipe	7/16	
3/8 Pipe	37/64	

TAPPING LUBRICANTS

Material	Drilling	Threading
Aluminum	Kerosene	Water Soluble Oil
Brass	Soluble Oil	Lard Oil
Bronze	Kerosene	Water Soluble Oil
Cast Iron	Water Soluble Oil	Lard Oil
Cast Steel	Water Soluble Oil	Lard Oil
Malleable Iron	Soda Water	Lard Oil
Monel Metal	Water Soluble Oil	Lard Oil
Stainless Steel	Kerosene	Lard Oil
Mild Steel	Water Soluble Oil	Lard Oil
Alloy Steel	Lard Oil	Lard Oil

Lubricants: Cutting lubricants reduce wear on taps and dies and keep the cutting edges cool. Water-soluble oil (a mineral oil base) is often used in shops, but home-based street rodders will find lard oil (Crisco and similar products), as well as kerosene, highly efficient cutting lubricants.

TECHNIQUES: NUT-SERTS TO YOU!

The chassis is the component to which many things are fastened, and with the exception of additional frame members that are most often welded in, a bolt or screw is usually needed. Bolts and screws, however, require an internally threaded base, and there are several tight places on the typical chassis where one or the other simply cannot be accommodated. Naturally, that particular place is the most appropriate for your add-on.

Of course, the chassis isn't the only location you may be eyeing. Recognizing that the most abused part of any street rod is the firewall, and though we may prefer to keep it accessory-free, occasionally we have to make an exception. At some future time, however, easy access to the cockpit side of the firewall is going to be restricted. Don't fret, there is an out.

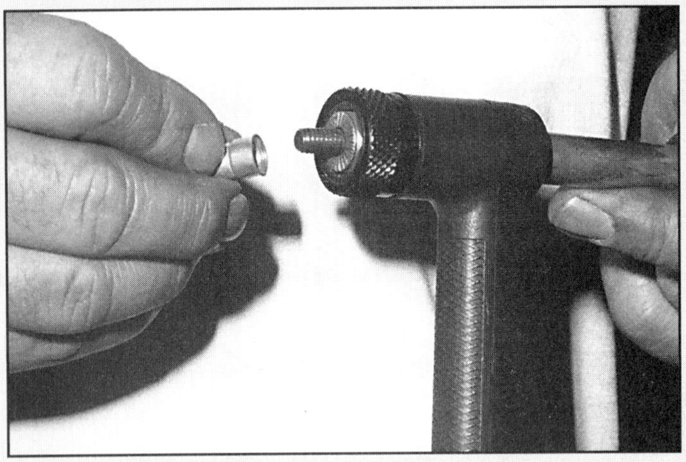

The most cost-effective way to place permanent threads in a blind spot is with a "rivet-nut" or "thread-insert" tool. There are several varieties of this type of tool for thin or thick materials. Use either name when you want to see what's available at your favorite industrial hardware retailer. Some feature both internal and external thread options.

The task at hand was the need for a permanent internal thread in Jim Bremner's chassis. The solution was a rivet-nut. After drilling a hole in the frame, the rivet nut was installed on the mandrel of the installation tool.

Depending on which brand of tool is chosen, the handle expands the rear of the rivet nut and locks it in place—ready for insertion of the fastener.

Neat, huh? Such threading resists high loads and doesn't loosen under vibration.

EPILOGUE

Several years ago, on a hot summer day, I walked into a convenience store to get a soft drink. As is my habit, I strolled over to the magazine rack to glance at the new issues. Two young men of about 19 were looking at the car mags and one picked up a copy of *Street Rodder*. The other turned to him and remarked, "That's for old men!" I'm sure my face went red as a beet, and I turned and walked out to my coupe and drove away. This anecdote is not apocryphal, it's the absolute truth. Since then I've mentioned it to a number of cohorts, and if memory serves, even to the editor of *SRM*.

I can't deny its reality, either. The vast majority of practicing street rodders is at least showing the early signs of male pattern baldness. And I am reasonably certain the vast majority of street rods display more resto-rod tradition than avant-garde styling. It's called nostalgia, and it runs rampant throughout our hobby.

And so, with that in mind, I'll close this book with a few photos from my archives. They were taken some years ago when Southern California's Gene Scott (the Godfather of the street rodding industry) still had his own personal surplus vintage tin inventory. Every time I've looked at the pictures, I am reminded of other wrecking yards, particularly the wrecking yards of my youth in the mid-1950s. Two of my favorites lay on the swampy outskirts of New Orleans. Although I don't remember their formal names, one was known as "Charlie's" and we called the other "The Fat Man's."

What I do remember, and will never forget, are the happy times scrounging through them with my pals looking for parts for my '40 coupe, or Robert Rider's '36 3-Window, or Lionel Duhon's '39 Chevy coupe.

Because you picked up this book, I'll wager you have your own lifetime of fond memories . . .

METRIC CUSTOMARY UNIT EQUIVALENTS

Multiply:	by:	to get:	Multiply by:	to get:
LINEAR				
inches	x 25.4	= millimeters (mm)	x 0.03937	= inches
feet	x 0.3048	= meters (m)	x 3.281	= feet
yards	x 0.9144	= meters (m)	x 1.0936	= yards
AREA				
inches^2	x 645.16	= millimeters^2(mm)	x 0.00155	= inches^2
feet^2	x 0.0929	= meters^2(m)	x 10.764	= feet^2
VOLUME				
quarts	x 0.94635	= liters (I)	x 1.0567	= quarts
gallons	x 3.7854	= liters (I)	x 0.2642	= gallons
feet^3	x 28.317	= liters (I)	x 0.03531	= feet^3
feet^3	x 0.02832	= meters^3(m^3)	x 35.315	= feet^3
fluid oz	x 29.57	= milliliters (ml)	x 0.03381	= fluid oz
MASS				
ounces (av)	x 28.35	= grams (g)	x 0.03527	= ounces (av)
pounds (av)	x 0.4536	= kilograms (kg)	x 2.2046	= pounds (av)
FORCE				
ounces-f(av)	x 0.278	= newtons (N)	x 3.597	= ounces-f(av)
pounds-f(av)	x 4.448	= newtons (N)	x 0.2248	= pounds-f(av)

TEMPERATURE

Degrees Celsius (C) = 0.556 (F – 32)Degree Fahrenheit (F) = (1.8C) + 32

ENERGY OR WORK (Watt-second = joule= newton-meter)				
foot-pounds	x 1.3558	= joules (J)	x 0.7376	= foot-pounds
Btu	x 1055	= joules (J)	x 0.000948	= Btu
PRESSURE OR STRESS				
pounds/sq in.	x 6.895	= kilopascals (kPa)	x 0.145	= pounds/sq in
TORQUE				
pound-inches	x 0.11298	= newton-meters (N-m)	x 8.851	= pound-inches
pound-feet	x 1.3558	= newton-meters (N-m)	x 0.7376	= pound-feet
pound-inches	x 0.0115	= kilogram-meters (Kg-M)	x 87	= pound-feet
pound-feet	x 0.138	= kilogram-meters (Kg-M)	x 7.25	= pound-feet
POWER				
horsepower	x 0.74570	= kilowatts (kW)	x 1.34102	= horsepower

COMMON METRIC PREFIXES

mega (M) = 1,000,000 or 106 centi(c) = 0.01 or 10-2
kilo (k) = 1,000 or 103 milli (m) = 0.001 or 10-3
hecto (h) = 100 or 102 micro (u) = (u)0.000,001 or 10-6

ABOUT THE AUTHOR

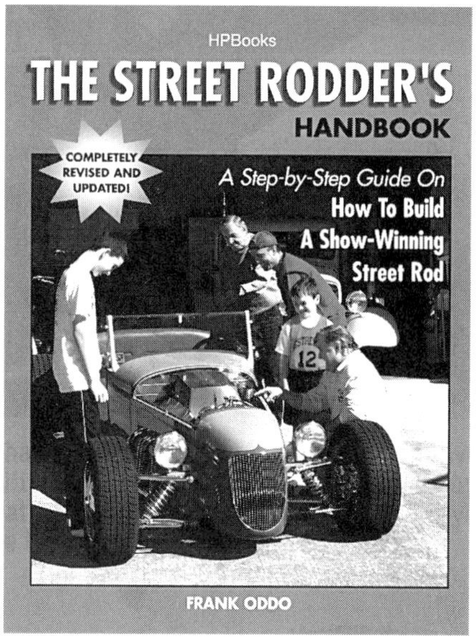

New Orleans born and bred Frank Oddo has been immersed in all forms of hot rodding for more than five decades. He started reading *Hot Rod Magazine* in grammar school two years before his dad taught him to drive. His second car, a 1940 Ford DeLuxe business coupe was purchased for $125 in 1955. Six months later, the Edmunds equipped 8BA flathead was pulled when he bought an overbored 1953 Cadillac short block. Once together, he equipped it with a mild Iskenderian cam and an Edelbrock staggered four (97) carbed manifold.

A few years and rods later, he bought his present coupe in 1968. Far and away the best bodied to date, it was powered by a Chevy 283, but backed up again, with a '39 box and stock rearend. A few weeks later he was in Harold Looney's fledgling early Ford parts store (which eventually became Vintique/C. W. Moss). There, he noticed a flyer on the bulletin board announcing the formation of a street rod club called "Forties Limited, Inc."

Needless to say, he became a charter member, and it wasn't long before he became editor of the club newsletter. In 1969 FLI assigned him the task of promoting the club by means of a magazine article. A two-pager with photos describing the club and its social activities was submitted to *Rod & Custom Magazine*, and much to everyone's delight, it was actually published!

Since then, he has publicly revealed his fascination with high-performance "old cars" in hundreds of articles and columns on the subject. Just one example of such prolific endeavor is the phrase "fat fendered" that is today commonly used in reference to 1938–'40 Fords. Oddo coined it in an article published in *Street Rodder Magazine* in 1972. Oddo was the first, if not the only, freelance photographer ever to gain permission to shoot street rods and custom trucks at Disneyland. Moreover, he did it three times.

He was on the Southern California Timing Association (SCTA) board of directors in the early 1970s and an associate editor and technical editor of *Street Rodder Magazine* throughout the 1970s.

Oddo not only writes about street rods (he has two other books in print, including HPBooks' *Street Rodder's Handbook*) he is a hot rodder through and through. He has built and raced roadsters on the Bonneville Salt Flats and El Mirage, the Southern California dry lake. Oddo held a Street Roadster record in the late '70s and a 225 mph Blown Gas Lakester record in the late 1990s. He still campaigns his Ford Boss 302-powered WWII P-38 belly tank lakester. Many such cars have come and gone since Bill Burke invented them in the 1940s, but Oddo's Cal'donia with its 25-year pedigree is the most senior single-owner belly tank participating on the straightaways.

Nevertheless, it is as "Mister Forty Ford" as *Hot Rod Magazine* once called him, that he is best known to his readership. In the enviable position of being a post-WWII street rodding participant almost from the beginning, and meeting many of the hot rod heroes of his youth, he has been on the scene when things happened, and has been able to record them for posterity!

GENERAL MOTORS
Big-Block Chevy Engine Buildups: 1-55788-484-6/HP1484
Big-Block Chevy Performance: 1-55788-216-9/HP1216
Camaro Performance: 1-55788-057-3/HP1057
Camaro Owner's Handbook ('67–'81): 1-55788-301-7/HP1301
Camaro Restoration Handbook ('67–'81): 0-89586-375-8/HP1375
Chevelle/El Camino Handbook: 1-55788-428-5/HP1428
Chevy S-10/GMC S-15 Handbook: 1-55788-353-X/HP1353
Chevy Trucks: 1-55788-340-8/HP1340
How to Hot Rod Big-Block Chevys: 0-912656-04-2/HP104
How to Hot Rod Small-Block Chevys: 0-912656-06-9/HP106
How to Rebuild Small-Block Chevy LT-1/LT-4: 1-55788-393-9/HP1393
John Lingenfelter: Modify Small-Block Chevy: 1-55788-238-X/HP1238
LS1/LS6 Small-Block Chevy Performance: 1-55788-407-2/HP1407
Powerglide Transmission Handbook:1-55788-355-6/HP1355
Rebuild Big-Block Chevy Engines: 0-89586-175-5/HP1175
Rebuild Gen V/Gen VI Big-Block Chevy: 1-55788-357-2/HP1357
Rebuild Small-Block Chevy Engines: 1-55788-029-8/HP1029
Small-Block Chevy Engine Buildups: 1-55788-400-5/HP1400
Small-Block Chevy Performance: 1-55788-253-3/HP1253
Turbo Hydramatic 350 Handbook: 0-89586-051-1/HP1051

FORD
Ford Windsor Small-Block Performance: 1-55788-323-8/HP1323
Mustang Performance (Engines): 1-55788-193-6/HP1193
Mustang Performance 2 (Chassis): 1-55788-202-9/HP1202
Mustang Performance Engine Tuning: 1-55788-387-4/HP1387
Mustang Restoration Handbook ('64–'70): 0-89586-402-9/HP1402
Rebuild Big-Block Ford Engines: 0-89586-070-8/HP1070
Rebuild Ford V-8 Engines: 0-89586-036-8/HP1036
Rebuild Small-Block Ford Engines: 0-912656-89-1/HP189

MOPAR
Big-Block Mopar Performance: 1-55788-302-5/HP1302
How to Hot Rod Small-Block Mopar Engine Revised: 1-55788-405-6/HP1405
How to Maintain & Repair Your Jeep: 1-55788-371-8/HP1371
How to Modify Your Jeep Chassis/Suspension for Offroad: 1-55788-424/HP1424
How to Modify Your Mopar Magnum V8: 1-55788-473-0/HP1473
How to Rebuild Your Mopar Magnum V8: 1-55788-431-5/HP1431
How to Rebuild and Modify The Chrysler 426 Hemi: 978-1-55788-525-8/HP1525
Rebuild Big-Block Mopar Engines: 1-55788-190-1/HP1190
Rebuild Small-Block Mopar Engines: 0-89586-128-3/HP1128
Torqueflite A-727 Transmission Handbook: 1-55788-399-8/HP1399

IMPORTS
Baja Bugs & Buggies: 0-89586-186-0/HP1186
Honda/Acura Engine Performance: 1-55788-384-X/HP1384
Honda/Acura Performance: 1-55788-384-X/HP1384
How to Hot Rod VW Engines: 0-912656-03-4/HP103
Mitsubishi & Diamond Star Performance Tuning: 978-1-55788-496-1/HP496
Porsche 911 Performance: 1-55788-489-7/HP489
Rebuild Air-Cooled VW Engines: 0-89586-225-5/HP1225
The VW Beetle: A History of The World's Most Popular Car: 1-55788-421-8/HP1421
The Volkswagen Super Beetle Handbook: 1-55788-483-8/HP1483

HANDBOOKS
Automotive Detailing: 1-55788-288-6/HP1288
Auto Electrical Handbook: 0-89586-238-7/HP1238
Auto Math Handbook: 1-55788-020-4/HP1020
Automotive Paint Handbook: 1-55788-291-6/HP1291
Auto Upholstery & Interiors: 1-55788-265-7/HP1265
Car Builder's Handbook: 1-55788-278-9/HP1278
Classic Car Restorer's Handbook: 1-55788-194-4/HP1194
Engine Builder's Handbook: 1-55788-245-2/HP1245
Engine Cooling Systems: 978-1-55788-425-1/HP1425
Fiberglass & Other Composite Materials Rev.: 1-55788-498-6/HP1498
The Lowrider's Handbook: 1-55788-383-1/HP1383
Metal Fabricator's Handbook: 0-89586-870-9/HP1870
1001 High Performance Tech Tips: 1-55788-199-5/HP1199
1001 MORE High Performance Tech Tips: 1-55788-429-3/HP1429
Paint & Body Handbook: 1-55788-082-4/HP1082
Performance Ignition Systems: 1-55788-306-8/HP1306
Pro Paint & Body: 1-55788-394-7/HP1394
Sheet Metal Handbook: 0-89586-757-5/HP1757
Welder's Handbook Revised: 978-1-55788-513-5/HP1513

INDUCTION
Holley 4150: 0-89586-047-3/HP1047
Holley Carbs, Manifolds & F.I.: 1-55788-052-2/HP1052
Rochester Carburetors: 0-89586-301-4/HP1301
Tuning Accel/DFI 6.0 Programmable F.I.: 1-55788-413-7/HP1413
Turbochargers: 0-89586-135-6/HP1135
Street Turbocharging: 1-55788-488-9/HP1488
Weber Carburetors: 0-89586-377-4/HP1377

RACING & CHASSIS
Bracket Racing: 1-55788-266-5/HP1266
Chassis Engineering: 1-55788-055-7/HP1055
4Wheel & Off-Road's Chassis & Suspension: 1-55788-406-4/HP1406
Dirt Track Chassis & Suspension: 978-1-55788-511-1/HP1511
How to Make Your Car Handle: 0-912656-46-8/HP146
How to Get Started in Stock Car Racing: 1-55788-468-4/HP1468
How to Build a Winning Drag Race Chassis & Suspension: 978-1-55788-462-6/HP1462
Racing Engine Builder's Handbook: 1-55788-492-7/HP1492
Stock Car Racing Engine Technology, 978-1-55788-506-7/HP1506
Stock Car Setup Secrets: 1-55788-401-3/HP1401

MOTORCYCLES
American V-Twin Engine Tech: 1-55788-455-2/HP1455

STREET RODS
How to Build a 1934–'35 Chevy Street Rod: 978-1-55788-514-1/HP1514
How to Build a 1932 Ford Street Rod: 1-55788-478-1/HP1478
How to Build a 1933–'34 Ford Street Rod: 1-55788-479-X/HP1479
How to Build a 1935–'40 Ford Street Rod: 1-55788-493-5/HP1493
Street Rodder magazine's Chassis & Suspension Handbook: 1-55788-346-7/HP1346
Street Rodder's Handbook, Rev.: 1-55788-409-9/HP1409
Street Rodding Tips & Techniques, 978-155788-515-9/HP1515

ORDER YOUR COPY TODAY!
All books can be purchased at your favorite retail or online bookstore such as amazon.com and barnesandnoble.com (use ISBN number), or auto parts store (use HP part number). You can also order direct from HPBooks by calling toll-free at 800-788-6262, ext. 1, or at www.penguin.com.